Advanced Imaging Applications for Interdisciplinary Engineering

Edited by

Ramesh Kumar
Department of Electronics and Communication Engineering
Manipal University Jaipur
Jaipur, Rajasthan
India

Anupma Gupta
Department of Interdisciplinary
Courses in Engineering, Chitkara University
Institute of Engineering and Technology
Chitkara University, Punjab
India

Manish Singla
Department of Interdisciplinary
Courses in Engineering, Chitkara University
Institute of Engineering and Technology
Chitkara University, Punjab
India

Arjun Puri
Department of Computer Science Engineering
Chitkara University Institute of Engineering and Technology
Chitkara University, Punjab
India

Ashish Kumar Singh
Departments of Electronics and Communication Engineering
Graphic Era (Deemed to be University)
Dehradun, Uttarakhand-248001
India

&

Vipan Kumar
Department of Electronics & Communication Engineering
Sri Sai College of Engineering & Technology
Punjab, India

Advanced Imaging Applications for Interdisciplinary Engineering

Editors: Ramesh Kumar, Anupma Gupta, Manish Singla, Arjun Puri, Ashish Kumar Singh & Vipan Kumar

ISBN (Online): 979-8-89881-456-4

ISBN (Print): 979-8-89881-457-1

ISBN (Paperback): 979-8-89881-458-8

Published by Bentham Science Publishers Pte. Ltd. Singapore, in collaboration with
Eureka Conferences, USA.

First published in 2026.

Bentham Science Publishers Pte. Ltd.
No. 9 Raffles Place
Office No. 26-01
Singapore 048619
Singapore
Email: subscriptions@benthamscience.net

CONTENTS

FOREWORD

It is with great enthusiasm that I introduce "Advanced Imaging Applications for Interdisciplinary Engineering." In this remarkable volume, the authors explore the ever-evolving landscape of imaging technology, highlighting its profound impact across multiple engineering domains. Once a supplementary tool, imaging has now become a driving force in enhancing visual interpretation and analytical capabilities fundamentally reshaping traditional methodologies and enabling novel breakthroughs. Charting the trajectory from early imaging techniques to today's sophisticated systems, the book offers a thorough account of the field's progression.

The authors skillfully build a strong conceptual foundation, making complex topics approachable for readers from varied disciplinary backgrounds. They systematically unpack the core principles and design strategies that define modern imaging systems. In parallel, the book delves into state-of-the-art simulation methodologies that have become essential for modeling and predicting system behavior in diverse applications. By leveraging these advanced computational approaches, engineers and researchers can navigate the complexities of imaging systems with greater precision, overcoming the constraints of conventional experimental techniques.

This narrative thoughtfully underscores how simulation serves as a catalyst for enhancing imaging device performance ultimately unlocking their untapped potential. As you turn these pages, I encourage you to immerse yourself in the depth of insight and expertise they contain. Whether you are an experienced professional, an emerging innovator, or simply intrigued by the frontiers of imaging science, this volume will undoubtedly expand your perspective and inspire you to advance the field.

Prabhat Singh
Department of Electronics and Communication Engineering
Engineering Indian Institute of Information Technology
Manipur, India

PREFACE

The field of engineering has always been driven by innovation and the ability to translate ideas into perceptible solutions. In recent years, one of the most transformative forces driving this innovation has been the advancement of imaging technologies. This book, Advancement of Imaging Application for Interdisciplinary Engineering, delves into this exciting intersection, exploring how cutting-edge imaging techniques are revolutionizing diverse engineering disciplines. Imaging applications in engineering have transcended their role in simply visualizing objects. They now offer powerful tools for non-destructive testing, material characterization, process monitoring, and design optimization. This book provides a comprehensive overview of these advancements, highlighting their impact on various engineering fields. Beyond the individual disciplines, this book underscores the importance of interdisciplinary collaboration. By fostering communication and knowledge exchange between imaging specialists and engineers, we can unlock the full potential of these technologies and accelerate innovation across diverse fields. Throughout the book, we explore the fundamental principles of various imaging techniques, delve into their practical applications in different engineering settings, and discuss emerging trends that promise to further revolutionize the field.

Ramesh Kumar
Department of Electronics and Communication Engineering
Manipal University Jaipur
Jaipur, Rajasthan
India

Anupma Gupta
Department of Interdisciplinary
Courses in Engineering, Chitkara University
Institute of Engineering and Technology
Chitkara University, Punjab
India

Manish Singla
Department of Interdisciplinary
Courses in Engineering, Chitkara University
Institute of Engineering and Technology
Chitkara University, Punjab
India

Arjun Puri
Department of Computer Science Engineering
Chitkara University Institute of Engineering and Technology
Chitkara University, Punjab
India

Ashish Kumar Singh
Departments of Electronics and Communication Engineering
Graphic Era (Deemed to be University)
Dehradun, Uttarakhand-248001
India

&

Vipan Kumar
Department of Electronics & Communication Engineering
Sri Sai College of Engineering & Technology
Punjab
India

List of Contributors

Anchal Sharma	Department of Civil Engineering, SSGI Pathankot, Punjab, India
Anupma Gupta	Department of Electronics and Communication Engineering, Chandigarh University, Mohali, Punjab, India
Arjan Singh	Department of Mechanical Engineering, CTIEMT, Punjab, India
Bh. Dasaradha Ram	CSE (AI&ML) department, NRI Institute of Technology, Andhra Pradesh-521212, India
Bhushan Dua	Department of Computer Science, Chandigarh University, Mohali, Punjab, India
D. V. Naga Raju	Department of Information Technology, Shri Vishnu Engineering College for Women, Bhimavaram, AP, India
Dhiraj Kumar Singh	, Galgotias University, Greater Noida, Uttar Pradesh-226001, India
Gaganpreet Kaur	Chitkara University Institute of Engineering and Technology, Chitkara University, Rajpura, Punjab-140401, India
Gajendra Singh	School of Engineering and Sciences, G. D. Goenka University, Gurugram-122103, India
Kanwarpreet Kaur	Department of Electronics and Communication Engineering, Thapar Institute of Engineering and Technology, Patiala, Punjab, India
Krishna Prakash	Department of Information and Communication Technology, Marwadi University, Rajkot, Gujarat-360003, India
Kumar Singh	Department of Electronics and Communication Engineering, Graphic Era (Deemed to be University), Dehradun, India
Lokesh Pawar	Amity School of Engineering and Technology, Amity University, Punjab, Mohali, India
M. Bhanurangarao	Department of Information Technology, Shri Vishnu Engineering College for Women, Bhimavaram, AP, India
M. Srikanth	Department of Information Technology, SRKR Engineering College, Bhimavaram, AP, India
Manish Kumar Singla	Department of Biosciences, Saveetha School of Engineering. Saveetha Institute of Medical and Technical Sciences, Chennai-602105, India
Manoj Kumar	School of Electrical and Electronics Engineering, Lovely Professional University, Phagwara, India
Meduri Raghuchandra	Department of Information Technology, Shri Vishnu Engineering College for Women, Bhimavaram, AP, India
Meet Kumari	ECE Department, National Institute of Technology, Delhi, India
Mohammad Aljaidi	Department of Computer Science, Zarqa University, Zarqa-13110, Jordan
Mohammed H. Alsharif	Department of Electrical Engineering, College of Electronics and Information Engineering, Sejong University, Gwangjin-gu, Seoul-05006, Korea

Muhammed Ali S.A.	Fuel Cell Institute, Universiti Kebangsaan Malaysia, Selangor 43600, Malaysia
Payal Patial	Department of Electronics and Communication Engineering, Chandigarh University, Mohali, Punjab, India
Prince Jain	Mechatronics Department, Parul University, Vadodara, Gujarat, India
Radhika Singla	Department of Computer Science, Chandigarh University, Mohali, Punjab, India
Ramesh Kumar	Department of Electronics and Communication Engineering, Manipal University Jaipur, Jaipur, Rajasthan, India
Ravina Gill	Department of Computer Science and Engineering, Chandigarh University, Punjab, India
Renuka Chowdary Bheemana	Department of Information and Communication Technology, Marwadi University, Rajkot, Gujarat-360003, India
Shonak Bansal	Department of Electronics and Communication Engineering, Chandigarh University, Mohali, Punjab, India
Spinder Kaur	Department of Computer Science and Engineering, Chandigarh University, Punjab, India
Taranjeet Kaur	Department of Computer Science, Chandigarh University, Mohali, Punjab, India
V. Lakshmikanth Chowdary	CSE (AI&ML) department, NRI Institute of Technology, Andhra Pradesh-521212, India
Vipan Kumar	Department of Electronics & Communication Engineering, Sri Sai College of Engineering & Technology, Punjab, India
Vishal Vishnoi	Department of Electronics and Communication Engineering, National Institute of Technology Sikkim, Sikkim, India
Yesu Jyothi	Department of Information Technology, Shri Vishnu Engineering College for Women, Bhimavaram, AP, India

CHAPTER 1

Investigation of the Imaging Algorithms in Artificial Intelligence

Kanwarpreet Kaur[1,*], Payal Patial[5], Shonak Bansal[3], Meet Kumari[2] and **Muhammed Ali S.A.[4]**

[1] *Department of Electronics and Communication Engineering, Thapar Institute of Engineering and Technology, Patiala, Punjab, India*

[2] *ECE Department, National Institute of Technology, Delhi, India*

[3] *Department of Electronics and Communication Engineering, Chandigarh University, Mohali, Punjab, India*

[4] *Fuel Cell Institute, Universiti Kebangsaan Malaysia, Selangor 43600, Malaysia*

[5] *Department of Electronics and Communication Engineering, Chandigarh University, Gharuan, Punjab, India*

Abstract: Artificial Intelligence (AI) is responsible for the transformation of image processing through cutting-edge imaging algorithms. This chapter delves deeper into AI-based imaging algorithms, such as image classification and pattern recognition. It provides a detailed review of the utilization of machine learning and deep learning models in imaging algorithms. Further, the applications of these AI-driven algorithms are also explored, thus emphasizing the advantages of these models in the real world. Key challenges and opportunities in AI-driven imaging are discussed, offering insights into emerging research directions.

Keywords: Deep learning, Feature engineering, Image classification, Machine learning.

INTRODUCTION

Imaging algorithms have always been indispensable for acquiring as well as exchanging information. These algorithms were earlier used for acquiring, transforming, restoring, enhancing, segmenting, and extracting edges of images, but now, with the inception of Artificial Intelligence (AI), there is a significant advancement in the technologies of image algorithms, such as image analysis, image classification, image generation, pattern recognition, and object detection. Despite being the older concept, AI was formalized only in the mid-twentieth

[*] **Corresponding author Kanwarpreet Kaur:** Department of Electronics and Communication Engineering, Thapar Institute of Engineering and Technology, Patiala, Punjab, India; E-mail: kpreet2392@gmail.com

Ramesh Kumar, Anupma Gupta, Manish Singla, Arjun Puri, Ashish Kumar Singh & Vipan Kumar (Eds.)

century. The term "Artificial Intelligence" was coined by John McCarthy in 1956. Imaging algorithms in AI involve computer methods, such as processing, analyzing, and interpreting visual information. In the present day, AI-driven imaging algorithms have led to several innovations from healthcare to entertainment. These breakthroughs have become possible only because of the availability of data and computational resources [1 - 3].

The vital components of AI that are involved in these imaging algorithms are the Machine Learning (ML) and Deep Learning (DL) methods. These algorithms are basically designed to learn the features present in images for performing the tasks of classification, generation, recognition, and detection. In the AI-driven imaging algorithms, the ML and DL models, especially Convolutional Neural Networks (CNNs), are widely employed in various applications. These learning methods are capable of identifying patterns and features for several applications in the image processing domain, which primarily involve image segmentation, image classification and recognition, image enhancement, and image generation [2, 4].

Image segmentation is basically the division of images into meaningful regions, thus making it quite easy to analyze the particular image regions. ML and DL models, such as CNN, AlexNet, and GoogleNet, have gained prominence in performing segmentation, which is important for diagnosis in medical imaging [5, 6]. These algorithms are capable of getting a detailed analysis by segmenting the regions of interest (ROI), thus enhancing the performance metrics. The classification of the images or recognizing the patterns is another field in which the ML and DL models are capable of performing the classification into different categories for the datasets. It can be employed in the majority of imaging applications, such as face recognition, forgery detection, disease detection, *etc.*, which are used in real-time scenarios [7 - 9].

Further, DL-based superresolution algorithms, such as CNN and residual neural networks, are used for performing the enhancement of images. It involves the enhancement of image resolution, thus making the details prominent. It is beneficial in the case of medical and remote sensing images, in which details are required for performing the analysis to make significant decisions [10, 11]. Further, Generative Adversarial Networks (GANs) are utilized for generating as well as enhancing images, such as the generation of defective images from non-defective images, and performing image-to-image translation for medical images [12 - 14].

Thus, ML and DL are extensively employed for imaging algorithms in several sectors to increase their efficacy. The capability of the imaging algorithms increases to interpret the data with higher accuracy, thus paving the way for advanced applications. This chapter delves deeper into the historical evolution of AI in imaging algorithms before discussing the ML and DL approaches in imaging algorithms. Afterward, the AI-driven imaging algorithms are explored in various sectors in the subsequent sections before concluding.

HISTORICAL PERSPECTIVE

In the past, the imaging algorithms were focused on simpler applications, such as edge detection and enhancement, based on mathematical approaches, such as transforms. The realization of imaging algorithms using AI dates back to the mid-twentieth century during the exploration of computer vision. Artificial Neural Networks (ANN) and ML models came into the picture at that time to find the optimal solutions for the problems [15, 16].

Towards the end of the twentieth century, learning-based approaches, such as Support Vector Machines (SVM), were introduced. These methods were based on the features considered by the individuals, thus limiting the flexibility. Further, a significant breakthrough was made with the development of the CNN model, which led to the automatic analysis and accurate detection of patterns. A CNN-based method was proposed for recognizing the handwritten characters, which was a solution to the real-world problem. Further, they proposed the Modified National Institute of Standards and Technology (MNIST) dataset for handwritten character recognition [16, 17].

With the advent of DL in the past few decades, particularly with the inception of AlexNet, VGG, ResNets, GoogleNet, *etc.*, there has been rapid advancement in the AI-based applications of image processing in several fields. It has further led to devising solutions for the problems of face recognition, object detection, and disease detection in real-time. In the last decade, the advent of Generative Adversarial Networks (GANs) has further increased the possibilities of more advanced methods not only for performing image analysis, synthesis, and enhancement but also for creating realistic images from scratch [16, 18].

Thus, the evolution of AI-driven imaging has led to a shift from rule-driven approaches to data-driven approaches based on learning. These significant developments have led to the expansion in the implementation of AI across multiple scenarios, ranging from healthcare to industry.

BUILDING BLOCKS OF AI IN IMAGING ALGORITHMS

In broader terms, AI usually refers to the approach that mimics the intelligence of humans. Traditionally, AI was based on two directions, namely, connectionism and computationalism. The former followed a bottom-up approach, which involved the biological neuron-based models. It was based on the emergence of intelligence from learning by experience. In contrast, the latter one does not involve any biological implementation. Instead, it is based on formal logic and reasoning. The former direction involves learning-based AI approaches that further comprise ML and DL. These data-driven approaches are broadly categorized into three learning styles: that is, the basic learning framework, the hybrid learning framework, and learning strategies, as illustrated in Fig. (**1**). If the model is trained using labeled data to perform the prediction, then it is supervised learning, which is further divided into classification and regression. Here, classification is the categorization of data into different categories, while regression refers to the prediction of continuous values. If the model is trained with the unlabeled data to develop clusters or segments based on similarity, then it is known as unsupervised learning. It is also referred to as self-organization. Reinforcement learning possesses the capability to learn by interacting with an environment. It is based on the concept of learning from feedback obtained by either rewarding or punishing the agents after performing the task in a manner similar to trial and error [1, 19].

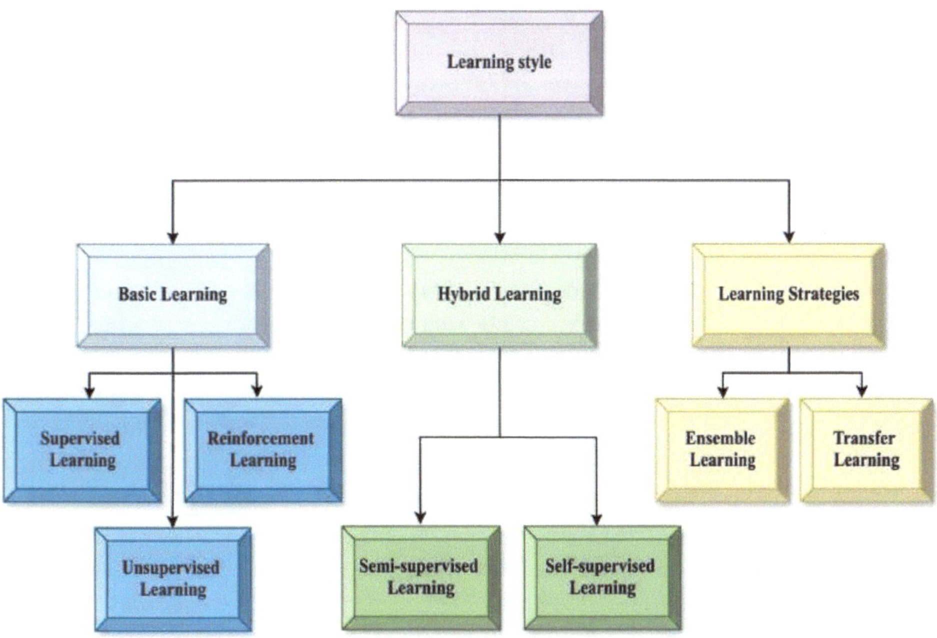

Fig (1). Learning framework categorization.

Apart from these categories, there are two hybrid learning frameworks. The first one lies amidst supervised and unsupervised learning, known as semi-supervised learning, for which the output is just partly known. The second one is self-supervised learning, which performs the task on the unlabelled data. The difference from unsupervised learning is that although the model is pre-trained on the unlabeled data, labels are generated automatically, which are used subsequently for performing the allocated task.

Further, the terms ensemble learning and transfer learning are also associated with the recent AI-based imaging applications. Ensemble learning is the combination of numerous models to enhance the model efficacy as well as the model stability. Transfer learning is based on the concept of utilizing previously trained models for different applications. In this, the performance of the model is enhanced by utilizing the existing models for different but relevant domains [1].

AI Imaging Algorithms

The AI involved in the imaging algorithms is basically the learning models, either ML or DL. This section briefly discusses some of the commonly used models. SVM is one of the statistics-based ML models used for performing classification into different categories by creating decision boundaries. In this model, a decision boundary, also known as a hyperplane, is created by the model in such a manner that the closest data points (support vectors) of two categories are as far as possible from each other. Originally, it was developed to perform linear classification, but it is also used as a kernel method for developing high-dimensional non-linear models. The non-linear problem is converted into a linear one by adding the extra dimensions to the considered data, thus resulting in a high-dimensional space. It is assistive in finding the patterns in a variety of applications [20].

Decision Trees (DTs) are widely applied for classification. It is composed of nodes and branches. Nodes are further divided into three types: the root node or decision node. This is a choice that performs the subdivision of data into two or more mutually exclusive sets. Internal nodes or chance nodes illustrate one of the available choices. From the top, it is connected to the parent node, while the bottom is connected to child or leaf nodes. Leaf nodes or end nodes provide the result as a combination of decisions. Branches are the occurrences that commence from either the root node or internal nodes. The path from the root to the leaf node *via* the internal node depicts the classification decision rule. The important steps of the DT model are as follows [21]:

1. Splitting: The target variable-related input variable is utilized to split the parent node into a child node. Initially, the most crucial input variable should be split at the root node and corresponding internal nodes into two or more categories, which are associated with certain features such as entropy, gain ratio, Gini index, information gain, *etc.* This process continues till the stopping criteria or homogeneity is satisfied.
2. Pruning: If the stopping criteria are not working in an effective manner, then a large tree is made based on the DT model, and then the nodes with less important information are removed.
3. Stopping: The stopping criterion is developed to prevent the model from becoming extremely complex.

Further, the ensemble of several DTs that are not related to each other is implemented for performing the classification in the case of Random Forests (RFs). The combination of various decisions of the DT can be used to attain the final result by either voting or internal feature selection. The mechanism is illustrated in Fig. (**2**) [1].

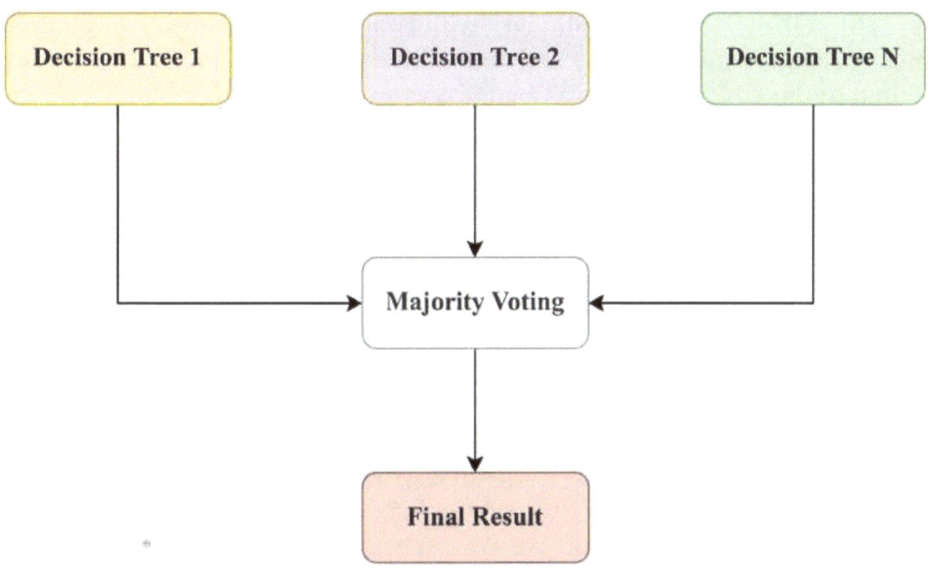

Fig. (2). Random Forest model based on the concept of majority voting.

Further, CNN models consist of several layers of convolutions and downsampling, as well as fully connected layers to obtain the outputs. The multiple convolutions allow the network to perform the extraction of simple features, such as edges. The fixed-size convolution kernel is applied to obtain the feature map. To reduce the size of these feature maps, downsampling layers are added between convolutional layers by applying the max-pooling operations. Finally, the fully connected layers are applied for generating the outputs [1].

GANs are the architectures that are used for generating outputs. It consists of a discriminator and a generator. In this, a discriminator is utilized for computing the distribution of original samples while the generator is employed for generating new samples from the original ones [22]. In the subsequent sections, the applicability of AI is explored in the various imaging algorithms.

AI in Image Segmentation

Image segmentation approaches have been widely utilized in medical image analysis. It is used for outlining the Regions of Interest (ROI) in the images, which can be further utilized for extracting the features for performing the disease diagnosis as well as other applications [23].

A fully convolutional network, U-Net, with a U-shaped architecture, is deployed for segmenting lung lesions and regions for COVID applications [23 - 25]. In a study [26], UNet++ was employed, which applies the nested convolutional architecture among the encoding and decoding paths. Although it is quite complex compared to the UNet, it provides better performance. Further, the V-Net is also applied, which utilizes the residual blocks as convolutional blocks, thus optimizing the network with the Dice loss [27]. Even the attention mechanisms are also employed for learning the discriminant features. Attention U-Net can capture fine details in medical images, enabling segmentation of lung nodules and lesions [23].

Image segmentation is achieved by employing the segnet-based encoder-decoder structure. This approach successfully attains pixel-level segmentation [28]. Mask R-CNN is also employed for the segmentation of images [29]. Further, the variants of GAN, such as Segmentation Adversarial Network, Structure Correcting Adversarial Network, Projective Adversarial Network, and Distributed Asynchronized Discriminator GAN, are utilized for the segmentation of medical images [30]. With the applicability of AI, especially DL, significant results can be attained in terms of speed as well as accuracy. Moreover, the evaluation metrics utilized for the segmentation are defined in Fig. (**3**).

Evaluation Metrics for Image Segmentation

Dice Index: It is utilized to evaluate the overlap or similarity among two samples. Its value lies between 0 and 1. Better segmentation results can be obtained if value is closer to 1.

Segmentation accuracy (SA): Area of accurate segmentation provides the percentage of real area in the Ground Truth image.

Hausdorff distance: It measures the degree of similarity between two sets of points which is basically the distance between boundaries of ground truth and the segmentation result input to the network.

Fig. (3). Evaluation metrics for Image Segmentation.

AI in Image Classification

In these AI-driven learning approaches, the whole task of classification revolves around the data, which can be raw data, augmented data, or processed data. The data for a specific task are divided into training and test sets based on certain criteria. The training data are specifically targeted for the machine to learn from, and the testing data are the unknown data used to evaluate the effectiveness of the algorithm. As imaging algorithms are considered, the data are in image form. The basic workflow of AI models is depicted in Fig. (**4**) [1].

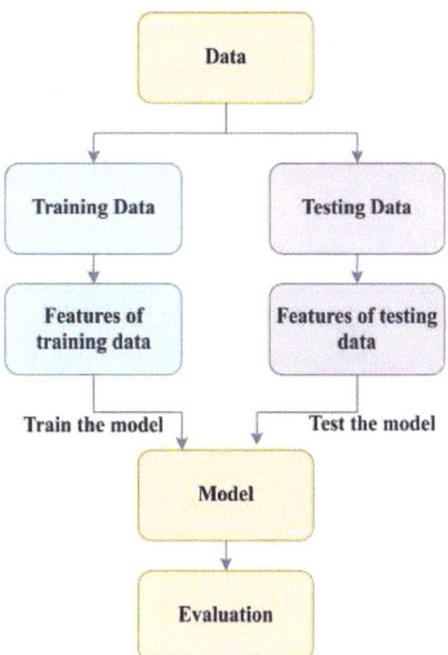

Fig. (4). Workflow of AI-based models.

As depicted in Fig. (**4**), initially, the data (images) are taken from a certain source, which is then split into training and testing data. Sometimes, images need to be pre-processed before splitting. Then, the features are extracted from the training data to train the model. When the model is trained, the features are extracted from the unknown data/testing data, which are then fed to the model for evaluating the model's performance. Eventually, the model's effectiveness is confirmed based on certain evaluation parameters. These are the basic steps involved in an ML-based workflow for imaging algorithms. In the case of DL, the features are extracted automatically from the images. Thus, the basic difference between the steps of ML and DL workflow can be considered as an evolution from feature engineering in ML to generic feature learning in DL.

In a study [31], the breast cancer classification was carried out by utilizing the DL and multiclass SVM. It involves the preprocessing approach before extracting the features by using K-means clustering, which obtained better results than DT. In another study [32], the concept of transfer learning was utilized for breast cancer classification, which applied the pre-trained VGG16 model. When compared to ResNet 50 and VGG19, it provided better performance for the same datasets. Similarly, classification methods are utilized for diagnosing cancer, arrhythmia, pneumonia, COVID-19, *etc.*, based on the considered dataset [33 - 35]. ML and DL-based classification is also employed in the case of image and video forgery detection. Several algorithms are applied after preprocessing the spliced or copy-move-based forged images to detect the tampered region in those images [36]. Further, Singular Value Decomposition (SVD) and Discrete Cosine Transformation (DCT) are employed to make the approach more robust against noise, transformations, and compression. Further, the classification is performed using SVM by extracting feature sets from preprocessed images. The images classified as forged use K-means clustering to localize the forged area [37]. Further, SVM is also employed for the detection of JPEG-compressed images by applying DCT [38].

Further, DL-based approaches are widely used in Natural Language Processing (NLP), particularly for sentiment analysis and text classification [39, 40]. Apart from this, learning-based methods are also employed for plant disease detection, where input images are fed, and classification is performed based on features extracted from them [41, 42]. The parameters used for classification are illustrated in Fig. (**5**).

The commonly used metrics are defined as follows:

1. Accuracy is determined by the count of correct predictions to the count of total predictions carried out in a particular example.

2. Precision is evaluated based on the count of the predictions that are true to the total positive predictions.
3. Recall is determined by computing the proportion of the actual positives that were predicted incorrectly.
4. F1-score is used for determining the performance of classifiers. It is computed using the harmonic mean of recall and precision.

Thus, AI is generally used for performing classification-based tasks, especially applications involving biomedical, industrial, and remote sensing.

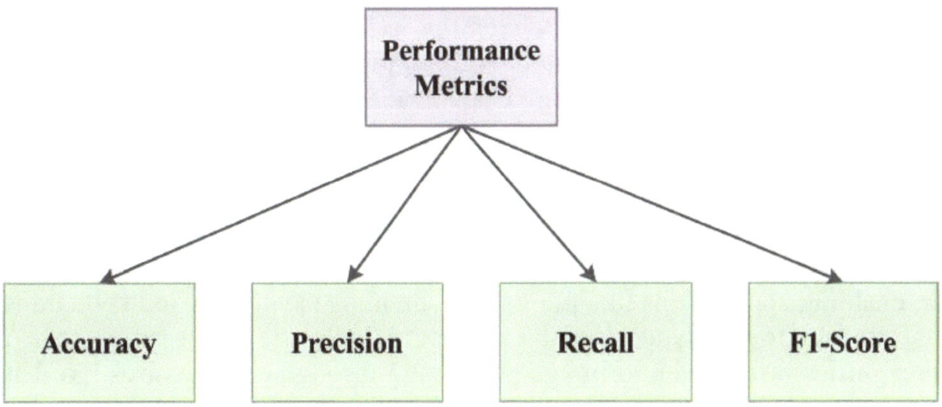

Fig. (5). Performance metrics for image classification algorithms.

AI in Image Enhancement

Image enhancement, which was previously carried out using the spatial domain or transform domain methods, is now improved using AI-driven techniques.

A CNN-based enhancement model is developed for the enhancement of low-light images [43]. In this, daytime images are transformed into low-light images by a generation pipeline, which is utilized for the construction of image pairs. The generated low-light images are used for training as well as validation of networks. In a study [44], low-light illumination methods based on unsupervised, supervised, semi-supervised, and zero-shot learning are studied and analyzed.

In a previous study [45], Magnetic Resonance Images (MRI) were of poor quality, which needed to be enhanced for further processing. In order to enhance those images, deep learning techniques were employed to improve the quality and correct the artefacts. Extending to other medical images, DL methods were employed to reduce speckle noise in the retinal images [46].

Another major area is underwater image enhancement, which uses deep inception and channel-wise attention modules to improve contrast, quality, and color in hazy images [47]. Apart from this, DL methods such as CNNs, GANs, and transformers are also applied to image super-resolution for reconstruction. There are different evaluation metrics for different methods, but the most commonly used are image-based comparisons with the original images or existing ones [48, 49].

DISCUSSION

As discussed, specific examples of machine learning and deep learning models, such as Convolutional Neural Networks (CNNs), Generative Adversarial Networks (GANs), and Transformers, are highlighted, emphasizing their unique contributions to image processing. CNNs, with their ability to automatically detect spatial hierarchies in images, have become the cornerstone of tasks such as object detection, facial recognition, and medical imaging analysis.

GANs, on the other hand, have revolutionized image generation and enhancement by pitting two neural networks—a generator and a discriminator—against each other, enabling applications like photo-realistic image synthesis and style transfer. Transformers, originally designed for natural language processing, have also made significant inroads into image processing through architectures like Vision Transformers (ViTs), which leverage self-attention mechanisms to capture long-range dependencies in visual data. By incorporating these models into the discussion, we provide a more comprehensive understanding of how advanced AI techniques are pushing the boundaries of what is possible in image processing, offering deeper insights into their technical nuances and real-world applications.

AI-driven imaging faces several significant challenges that can impact its effectiveness and scalability. One major issue is data bias, where imbalanced or unrepresentative datasets lead to skewed model performance, particularly for underrepresented groups. This can result in unfair or inaccurate outcomes, such as misdiagnoses in medical imaging. To mitigate this, strategies like diversifying datasets, employing fairness-aware algorithms, and conducting regular bias audits are essential. Another challenge is the computational cost associated with training and deploying advanced models like CNNs and Transformers, which require substantial processing power and energy. Techniques, such as model pruning, quantization, and the use of efficient architectures, such as MobileNet, can help reduce computational demands.

Additionally, the need for large annotated datasets poses a barrier, as acquiring and labeling data is time-consuming and expensive. Semi-supervised learning, transfer learning, and synthetic data generation using GANs are promising approaches to address this limitation. By addressing these challenges through innovative mitigation strategies, the field of AI-driven imaging can achieve more robust, equitable, and scalable solutions.

CHALLENGES AND OPPORTUNITIES

Despite its remarkable progress, AI faces several challenges that hinder its full potential. One major concern is data quality and availability, as AI systems rely heavily on large, high-quality datasets for training. Bias in data and algorithms can lead to unfair or discriminatory outcomes, raising ethical and social concerns. Another challenge is the "black box" nature of many AI models, which makes it difficult to understand and explain their decision-making processes. Computational requirements for training advanced AI models are also substantial, leading to environmental concerns due to high energy consumption. Additionally, the rapid pace of AI development has outpaced the creation of regulatory frameworks, resulting in uncertainties around accountability, privacy, and security. Addressing these challenges is crucial to ensuring the responsible and equitable deployment of AI technologies. Based on the study conducted, the key challenges for the AI-based imaging algorithms are as follows:

1. Large Datasets: In order to perform effectively, AI-driven algorithms need to be trained on large and high-quality datasets to capture the minute details present in the images that might be useful in certain applications.
2. Computational Cost: Training models is computationally intensive because they require training on large, complex datasets. Even the models are complex in some cases.
3. Transparency: Especially in the case of DL models, it is quite difficult to understand what is happening inside the model. In some of the works conducted, DL algorithms are also considered black boxes, as one cannot interpret the functioning of the model.

However, there are immense opportunities in the field of AI-driven imaging, driven by improvements in the accuracy and efficacy of systems, especially in pattern recognition, object detection, and disease classification. The continuous development of AI-driven algorithms can lead to robust and flexible systems in diverse fields.

CONCLUSION

To conclude, rapid advancements in AI have significantly impacted imaging algorithms, including image segmentation, classification, and enhancement. The study highlights the extensive use of machine learning and deep learning models in real-world scenarios. Further, the challenges as well as opportunities are discussed that provide future insights into the AI-driven fields. With the evolution of AI, new avenues are certainly open to revolutionize the technologies based on imaging algorithms.

FUTURE DIRECTIONS

The future of AI holds immense promise, with ongoing research and innovation paving the way for groundbreaking advancements. One key area of focus is the development of explainable AI (XAI), which aims to make AI systems more transparent and interpretable. Another emerging trend is the integration of AI with other technologies, such as the Internet of Things (IoT) and quantum computing, to create more powerful and efficient solutions. Researchers are also exploring the potential of general AI, which seeks to build systems capable of performing any intellectual task that a human can do. Ethical AI development will remain a priority, with efforts to establish global standards and guidelines for responsible use. As AI continues to evolve, its impact on society, economy, and culture will undoubtedly grow, shaping the future in ways we are only beginning to imagine.

REFERENCES

[1] A. Barragán-Montero, U. Javaid, G. Valdés, D. Nguyen, P. Desbordes, B. Macq, S. Willems, L. Vandewinckele, M. Holmström, F. Löfman, S. Michiels, K. Souris, E. Sterpin, and J.A. Lee, "Artificial intelligence and machine learning for medical imaging: A technology review", *Phys. Med.,* vol. 83, pp. 242-256, 2021.
[http://dx.doi.org/10.1016/j.ejmp.2021.04.016] [PMID: 33979715]

[2] D.J. Lin, P.M. Johnson, F. Knoll, and Y.W. Lui, "Artificial intelligence for MR image reconstruction: an overview for clinicians", *J. Magn. Reson. Imaging,* vol. 53, no. 4, pp. 1015-1028, 2021.
[http://dx.doi.org/10.1002/jmri.27078] [PMID: 32048372]

[3] X. Zhang, and W. Dahu, "Application of artificial intelligence algorithms in image processing", *J. Vis. Commun. Image Represent.,* vol. 61, pp. 42-49, 2019.
[http://dx.doi.org/10.1016/j.jvcir.2019.03.004]

[4] D. Dey, P.J. Slomka, P. Leeson, D. Comaniciu, S. Shrestha, P.P. Sengupta, and T.H. Marwick, "Artificial intelligence in cardiovascular imaging: JACC state-of-the-art review", *J. Am. Coll. Cardiol.,* vol. 73, no. 11, pp. 1317-1335, 2019.
[http://dx.doi.org/10.1016/j.jacc.2018.12.054] [PMID: 30898208]

[5] K. Kamnitsas, C. Ledig, V.F.J. Newcombe, J.P. Simpson, A.D. Kane, D.K. Menon, D. Rueckert, and B. Glocker, "Efficient multi-scale 3D CNN with fully connected CRF for accurate brain lesion segmentation", *Med. Image Anal.,* vol. 36, pp. 61-78, 2017.
[http://dx.doi.org/10.1016/j.media.2016.10.004] [PMID: 27865153]

[6] H.C. Shin, H.R. Roth, M. Gao, L. Lu, Z. Xu, I. Nogues, J. Yao, D. Mollura, and R.M. Summers, "Deep convolutional neural networks for computer-aided detection: CNN architectures, dataset characteristics and transfer learning", *IEEE Trans. Med. Imaging,* vol. 35, no. 5, pp. 1285-1298, 2016.
[http://dx.doi.org/10.1109/TMI.2016.2528162] [PMID: 26886976]

[7] N. Kaur, N. Jindal, and K. Singh, "A passive approach for the detection of splicing forgery in digital images", *Multimedia Tools Appl.,* vol. 79, no. 43-44, pp. 32037-32063, 2020.
[http://dx.doi.org/10.1007/s11042-020-09275-w]

[8] N. Sharma, R. Sharma, and N. Jindal, "Face-based age and gender estimation using improved convolutional neural network approach", *Wirel. Pers. Commun.,* vol. 124, no. 4, pp. 3035-3054, 2022. [http://dx.doi.org/10.1007/s11277-022-09501-8]

[9] T. Bansal, S. Gupta, and N. Jindal, "Deep learning-based comprehensive review on pulmonary tuberculosis", *Neural Comput. Appl.,* vol. 36, no. 12, pp. 6513-6530, 2024. [http://dx.doi.org/10.1007/s00521-023-09381-4]

[10] Y. Li, B. Sixou, and F. Peyrin, "A review of the deep learning methods for medical images super resolution problems", *IRBM,* vol. 42, no. 2, pp. 120-133, 2021. [http://dx.doi.org/10.1016/j.irbm.2020.08.004]

[11] P. Wang, B. Bayram, and E. Sertel, "A comprehensive review on deep learning based remote sensing image super-resolution methods", *Earth Sci. Rev.,* vol. 232, 2022.104110 [http://dx.doi.org/10.1016/j.earscirev.2022.104110]

[12] S. Niu, B. Li, X. Wang, and H. Lin, "Defect image sample generation with GAN for improving defect recognition", *IEEE Trans. Autom. Sci. Eng.,* vol. 17, no. 3, pp. 1-12, 2020. [http://dx.doi.org/10.1109/TASE.2020.2967415]

[13] A. You, J.K. Kim, I.H. Ryu, and T.K. Yoo, "Application of generative adversarial networks (GAN) for ophthalmology image domains: a survey", *Eye Vis. (Lond.),* vol. 9, no. 1, p. 6, 2022. [http://dx.doi.org/10.1186/s40662-022-00277-3] [PMID: 35109930]

[14] S. Lin, F. Qin, Y. Li, R.A. Bly, K.S. Moe, and B. Hannaford, "Lc-gan: Image-to-image translation based on generative adversarial network for endoscopic images", *Proc. 2020 IEEE/RSJ Int. Conf. Intelligent Robots Systems (IROS),* 2020pp. 2914-2920 [http://dx.doi.org/10.1109/IROS45743.2020.9341556]

[15] R.C. Gonzalez, and R.E. Woods, *Digital image processing.* Pearson education india: India, 2009.

[16] "History and evolution of machine learning: A timeline", [Online] Available on: https://www.techtarget.com/whatis/A-Timeline-of-Machine-Learning-History [Accessed July 30, 2024].

[17] Y. Lecun, L. Bottou, Y. Bengio, and P. Haffner, "Gradient-based learning applied to document recognition", *Proc. IEEE,* vol. 86, no. 11, pp. 2278-2324, 1998. [http://dx.doi.org/10.1109/5.726791]

[18] M.Z. Alom, "The history began from alexnet: A comprehensive survey on deep learning approaches", [Online] Available on https://arxiv.org/abs/1803.01164 [Accessed Aug. 6, 2024].

[19] E.F. Morales, and H.J. Escalante, A brief introduction to supervised, unsupervised, and reinforcement learning.*Biosignal processing and classification using computational learning and intelligence.* Academic Press, 2022, pp. 111-129. [http://dx.doi.org/10.1016/B978-0-12-820125-1.00017-8]

[20] S. Huang, N. Cai, P.P. Pacheco, S. Narrandes, Y. Wang, and W. Xu, "Applications of support vector machine (SVM) learning in cancer genomics", *Cancer Genomics Proteomics,* vol. 15, no. 1, pp. 41-51, 2018. [PMID: 29275361]

[21] Y.Y. Song, and Y. Lu, "Decision tree methods: applications for classification and prediction", *Shanghai Jingshen Yixue,* vol. 27, no. 2, pp. 130-135, 2015. [PMID: 26120265]

[22] M. Durgadevi, "Generative Adversarial Network (GAN): A general review on different variants of GAN and applications", in 2021 6th International Conference on Communication and Electronics Systems (ICCES), July 2021, pp. 1-8.

[23] F. Shi, J. Wang, J. Shi, Z. Wu, Q. Wang, Z. Tang, K. He, Y. Shi, and D. Shen, "Review of artificial intelligence techniques in imaging data acquisition, segmentation, and diagnosis for COVID-19",

IEEE Rev. Biomed. Eng., vol. 14, pp. 4-15, 2021.
[http://dx.doi.org/10.1109/RBME.2020.2987975] [PMID: 32305937]

[24] Y. Cao, Z. Xu, J. Feng, C. Jin, X. Han, H. Wu, and H. Shi, "Longitudinal assessment of COVID-19 using a deep learning–based quantitative CT pipeline: illustration of two cases", *Radiol. Cardiothorac. Imaging,* vol. 2, no. 2, 2020.e200082
[http://dx.doi.org/10.1148/ryct.2020200082] [PMID: 33778563]

[25] L. Huang, R. Han, T. Ai, P. Yu, H. Kang, Q. Tao, and L. Xia, "Serial quantitative chest CT assessment of COVID-19: a deep learning approach", *Radiol. Cardiothorac. Imaging,* vol. 2, no. 2, 2020.e200075
[http://dx.doi.org/10.1148/ryct.2020200075] [PMID: 33778562]

[26] R. Kumar, R.K. Ratnesh, J. Singh, R. Chandra, G. Singh, and V. Vishnoi, "Recent Prospects of Medical Imaging and Sensing Technologies Based on Electrical Impedance Data Acquisition System", *J. Electrochem. Soc.*, vol. 170, no. 11, 2023.117507
[http://dx.doi.org/10.1149/1945-7111/ad050f]

[27] F. Milletari, N. Navab, and S.A. Ahmadi, "V-net: Fully convolutional neural networks for volumetric medical image segmentation", in Proc. 4th Int. Conf. 3D Vision, Oct. 2016, pp. 565–571.
[http://dx.doi.org/10.1109/3DV.2016.79]

[28] V. Badrinarayanan, A. Kendall, and R. Cipolla, "Segnet: A deep convolutional encoder-decoder architecture for image segmentation", *IEEE Trans. Pattern Anal. Mach. Intell.,* vol. 39, no. 12, pp. 2481-2495, 2017.
[http://dx.doi.org/10.1109/TPAMI.2016.2644615] [PMID: 28060704]

[29] S. Ren, K. He, R. Girshick, and J. Sun, "Faster R-CNN: Towards real-time object detection with region proposal networks", *IEEE Trans. Pattern Anal. Mach. Intell.,* vol. 39, no. 6, pp. 1137-1149, 2017.
[http://dx.doi.org/10.1109/TPAMI.2016.2577031] [PMID: 27295650]

[30] X. Liu, L. Song, S. Liu, and Y. Zhang, "A review of deep-learning-based medical image segmentation methods", *Sustainability (Basel),* vol. 13, no. 3, p. 1224, 2021.
[http://dx.doi.org/10.3390/su13031224]

[31] P. Kaur, G. Singh, and P. Kaur, "Intellectual detection and validation of automated mammogram breast cancer images by multi-class SVM using deep learning classification", *Inform. Med. Unlocked,* vol. 16, 2019.100151
[http://dx.doi.org/10.1016/j.imu.2019.01.001]

[32] N. Rani, D.K. Gupta, and S. Singh, "Multi-class classification of breast cancer abnormality using transfer learning", *Multimedia Tools Appl.,* vol. 83, no. 30, pp. 75085-75100, 2024.
[http://dx.doi.org/10.1007/s11042-023-17832-2]

[33] Z. Ebrahimi, M. Loni, M. Daneshtalab, and A. Gharehbaghi, "A review on deep learning methods for ECG arrhythmia classification", *Expert Systems with Applications,* vol. 7, p. 100033, 2020.
[http://dx.doi.org/10.1016/j.eswax.2020.100033]

[34] A. Raza, H. Ayub, J.A. Khan, I. Ahmad, A. Salama, Y.I. Daradkeh, D. Javeed, A. Ur Rehman, and H. Hamam, "A hybrid deep learning-based approach for brain tumor classification", *Electronics (Basel),* vol. 11, no. 7, p. 1146, 2022.
[http://dx.doi.org/10.3390/electronics11071146]

[35] D.M. Ibrahim, N.M. Elshennawy, and A.M. Sarhan, "Deep-chest: Multi-classification deep learning model for diagnosing COVID-19, pneumonia, and lung cancer chest diseases", *Comput. Biol. Med.,* vol. 132, 2021.104348
[http://dx.doi.org/10.1016/j.compbiomed.2021.104348] [PMID: 33774272]

[36] N. Kaur, N. Jindal, and K. Singh, "Passive Image Forgery Detection Techniques: A Review, Challenges, and Future Directions", *Wirel. Pers. Commun.,* vol. 134, no. 3, pp. 1491-1529, 2024.
[http://dx.doi.org/10.1007/s11277-024-10959-x]

[37] G.S. Priyanka, G. Singh, and K. Singh, "An improved block based copy-move forgery detection technique", *Multimedia Tools Appl.,* vol. 79, no. 19-20, pp. 13011-13035, 2020. [http://dx.doi.org/10.1007/s11042-019-08354-x]

[38] A. Kumar, G. Singh, A. Kansal, and K. Singh, "Digital image forensic approach to counter the JPEG anti-forensic attacks", *IEEE Access,* vol. 9, pp. 4364-4375, 2021. [http://dx.doi.org/10.1109/ACCESS.2020.3048246]

[39] M.U. Salur, and I. Aydin, "A novel hybrid deep learning model for sentiment classification", *IEEE Access,* vol. 8, pp. 58080-58093, 2020. [http://dx.doi.org/10.1109/ACCESS.2020.2982538]

[40] A. Gasparetto, M. Marcuzzo, A. Zangari, and A. Albarelli, "A survey on text classification algorithms: From text to predictions", *Information (Basel),* vol. 13, no. 2, p. 83, 2022. [http://dx.doi.org/10.3390/info13020083]

[41] Ü. Atila, M. Uçar, K. Akyol, and E. Uçar, "Plant leaf disease classification using EfficientNet deep learning model", *Ecol. Inform.,* vol. 61, 2021.101182 [http://dx.doi.org/10.1016/j.ecoinf.2020.101182]

[42] W. Shafik, A. Tufail, A. Namoun, L.C. De Silva, and R.A.A.H.M. Apong, "A systematic literature review on plant disease detection: Motivations, classification techniques, datasets, challenges, and future trends", *IEEE Access,* vol. 11, pp. 59174-59203, 2023. [http://dx.doi.org/10.1109/ACCESS.2023.3284760]

[43] G. Li, Y. Yang, X. Qu, D. Cao, and K. Li, "A deep learning based image enhancement approach for autonomous driving at night", *Knowl. Base. Syst.,* vol. 213, 2021.106617 [http://dx.doi.org/10.1016/j.knosys.2020.106617]

[44] H. Tang, H. Zhu, L. Fei, T. Wang, Y. Cao, and C. Xie, "Low-illumination image enhancement based on deep learning techniques: a brief review", *Photonics,* vol. 10, no. 2, p. 198, 2023. [http://dx.doi.org/10.3390/photonics10020198]

[45] Z. Chen, K. Pawar, M. Ekanayake, C. Pain, S. Zhong, and G.F. Egan, "Deep learning for image enhancement and correction in magnetic resonance imaging—state-of-the-art and challenges", *J. Digit. Imaging,* vol. 36, no. 1, pp. 204-230, 2022. [http://dx.doi.org/10.1007/s10278-022-00721-9] [PMID: 36323914]

[46] K.J. Halupka, B.J. Antony, M.H. Lee, K.A. Lucy, R.S. Rai, H. Ishikawa, G. Wollstein, J.S. Schuman, and R. Garnavi, "Retinal optical coherence tomography image enhancement via deep learning", *Biomed. Opt. Express,* vol. 9, no. 12, pp. 6205-6221, 2018. [http://dx.doi.org/10.1364/BOE.9.006205] [PMID: 31065423]

[47] H.F. Tolie, J. Ren, and E. Elyan, "DICAM: Deep Inception and Channel-wise Attention Modules for underwater image enhancement", *Neurocomputing,* vol. 584, 2024.127585 [http://dx.doi.org/10.1016/j.neucom.2024.127585]

[48] R. Kumar, R.K. Ratnesh, J. Singh, A. Kumar, and R. Chandra, "IoT-Driven Experimental Framework for Advancing Electrical Impedance Tomography", *ECS J. Solid State Sci. Technol.,* vol. 13, no. 2, 2024.027002 [http://dx.doi.org/10.1149/2162-8777/ad2331]

[49] A. Garg, R.K. Ratnesh, R.K. Chauhan, N. Mittal, and H. Shankar, "Current Advancement and Progress in BioFET: A Review", *International Conference on Signal and Information Processing (IConSIP),* 2022pp. 1-7 [http://dx.doi.org/10.1109/ICoNSIP49665.2022.10007517]

CHAPTER 2

Broadband Photodetection through Few-Layer Graphene/ZnO/Si Dual-Heterojunction and its Comparative Study with Machine Learning

Shonak Bansal[1,*], **Krishna Prakash**[2], **Anupma Gupta**[1], **Meet Kumari**[3], **Payal Patial**[1], **Kanwarpreet Kaur**[4], **Lokesh Pawar**[5] and **Gaganpreet Kaur**[6]

[1] *Department of Electronics and Communication Engineering, Chandigarh University, Mohali, Punjab, India*

[2] *Department of Information and Communication Technology, Marwadi University, Rajkot, Gujarat-360003, India*

[3] *ECE Department, National Institute of Technology, Delhi, India*

[4] *Department of Electronics and Communication Engineering, Thapar Institute of Engineering and Technology, Patiala, Punjab, India*

[5] *Amity School of Engineering and Technology, Amity University, Punjab, Mohali, India*

[6] *Chitkara University Institute of Engineering and Technology, Chitkara University, Rajpura, Punjab-140401, India*

Abstract: The pursuit of highly efficient photodetectors has garnered substantial interest recently, driven by their wide-ranging applications spanning environmental monitoring, communication systems, and imaging technologies. This chapter examines how combining few-layer graphene (FLG), zinc oxide (ZnO), and silicon (Si) in a dual heterojunction design can create a photodetector that captures light across a wide range, from ultraviolet (UV) to near-infrared wavelengths. Utilizing the exceptional properties of graphene, the wide-bandgap nature of ZnO, and the scalability of Si as a substrate, the dual-heterojunction is optimized for UV-to-NIR light absorption and efficient carrier transport. A comprehensive simulation and analysis of the photodetector's characteristic parameters, including biased and unbiased operation, is conducted using the Silvaco Atlas TCAD software. The results show enhanced performance, with a least dark current density (Jdark) of 2.7×10–15 A/cm2, superior photocurrent density (Jlight) of 0.26 μA/cm2, along with a remarkable Jlight/Jdark ratio of 9.77×107, a 3-dB cut-off frequency of 7.36 THz, and a rapid rise (fall) time of 0.47 (0.88) ns at –1.0 V. Under illumination conditions, the photodetector exhibits a peak external quantum efficiency of 69.3%, photocurrent responsivity of 0.26 A/W, detectivity of 8.12×1015 cmHz1/2/W, and noise equivalent power of 7.79×10–20 W at –1.0 V bias. This

* **Corresponding author Shonak Bansal:** Department of Electronics and Communication Engineering, Chandigarh University, Mohali, Punjab, India; E-mail: shonakk@gmail.com

Ramesh Kumar, Anupma Gupta, Manish Singla, Arjun Puri, Ashish Kumar Singh & Vipan Kumar (Eds.)

research advances photodetection capabilities by demonstrating that combining p^+-FLG, n-ZnO, and n-Si in a dual-heterojunction design enables effective light detection from UV-to-NIR wavelengths. The findings suggest this integrated approach could lead to improved sensors and imaging devices. Moreover, a comprehensive comparative analysis is performed, utilizing the power of different machine learning (ML) regression models to evaluate their predictive performance in modeling the photodetector's behavior. The main goal is to determine the effectiveness of each ML regression model in accurately forecasting the photodetector's characteristics. This will be achieved by evaluating model performance across varying test set sizes using a range of metrics such as mean squared errors, root mean squared errors, mean absolute errors, and R^2 scores. The findings highlight the utility of machine learning algorithms in modeling and optimizing optoelectronic devices, offering valuable insights into their potential applications in advancing photodetector research and development endeavors.

Keywords: Broadband, Dark current density, Dual-heterojunction, Few-layer graphene, Machine learning, Near-infrared, Noise current, Photocurrent density, Quantum efficiency, Ultraviolet, ZnO.

INTRODUCTION

Photodetectors, also called photosensors, are critical in optoelectronic applications by transforming optical energy into electrical signals. To fulfill the diverse demands of applications like space communication, military surveillance, imaging, optical communications, and environmental sensing, cutting-edge photodetectors are being developed. This involves leveraging advanced materials and pioneering device architectures. Conventional photodetectors are constrained by a narrow detection bandwidth, restricting their capabilities. On the other hand, broadband photodetectors can employ a single device design to sense light spanning the ultraviolet (UV), visible, and infrared (IR) spectrums. Enhancing a photodetector's detection range substantially widens the scope of applications where it can be utilized.

Numerous efforts have been made to develop single broadband photodetectors by different research groups [1 - 4]. Despite their capabilities, the photodetectors developed [1 - 4] face limitations in terms of narrow detection ranges and relatively poor response times [5]. Moreover, attaining high photocurrent responsivity and stability in the UV spectral range, similar to the visible and IR ranges, while maintaining a fast response time continues to be difficult [6]. Silicon (Si) is the favored choice for manufacturing photodetectors sensitive to visible and near-infrared (NIR) wavelengths due to several advantageous factors. Firstly, the well-established technology surrounding Si manufacturing processes ensures a mature and reliable production framework. Additionally, Si's widespread availability contributes to its cost-effectiveness, making it an economically viable option for large-scale production. Furthermore, the manufacturing methods associated with Si are optimized for efficient and cost-effective fabrication, further reinforcing its position as the preferred material for photodetectors operating within the visible and NIR spectral ranges [5, 7]. Still, Si-based photodetectors face limitations, including high noise current, low photocurrent responsivity (less than 0.1 A/W for wavelengths below 400 nm), low efficiency, and a restricted spectral range due to the Si's narrow bandgap (~1.12 eV), sensitivity to temperature variations, high reflection coefficient, and low absorption [6, 8 - 10].

To address these shortcomings, the formation of a heterojunction between Si and wide-bandgap semiconductor materials has been proposed as a promising approach [10]. Among the various wide-bandgap semiconductors investigated [11], Zinc oxide (ZnO) emerges as a highly attractive material among wide-bandgap semiconductors due to its exceptional properties [11]. These include a large energy bandgap, better optical transparency in the visible spectrum region, substantial exciton binding energy, strong absorption in the UV spectrum, low material cost, and excellent stability at room temperature. ZnO's intrinsic n-type character offers a significant advantage, allowing for the straightforward fabrication of direct p-n heterojunctions by combining it with p-type materials [10, 12, 13]. ZnO can be effectively synthesized using a variety of physical and chemical vapor-phase deposition methods [10, 12, 14]. The heterojunction formed by integrating ZnO with Si emerges as a favorable option for optoelectronic devices, owing to its low fabrication costs, capability for low-temperature deposition, and superior optoelectronic properties. This ZnO/Si heterojunction exhibits the ability to detect radiation spanning the UV-to-NIR spectral regions within a single device framework [10, 12, 15], hence reducing the cost and size of the photodetectors.

Among the various two-dimensional (2D) materials explored so far for photodetection, graphene has gathered significant attention due to its exceptional properties. These properties include high sensitivity, carrier mobility, current handling capacity, low resistivity, broadband light absorption, and adjustable Fermi level [11, 16]. Consequently, graphene has gained considerable attention as a promising candidate for forming effective heterojunctions with other semiconductor materials in optoelectronic devices [10, 17 - 24]. The strong built-in electric field in these heterojunctions efficiently separates light-generated charges without needing an external voltage. This self-powered capability is key for developing high-performing photodetectors [10, 17 - 21, 23, 24]. The graphene layer plays a pivotal role by facilitating efficient carrier separation, attributed to its adjustable Fermi level [16]. Additionally, it functions as an antireflection coating, significantly reducing light reflection in the visible and NIR regions [25]. Therefore, integrating graphene with ZnO and Si in a dual-heterojunction configuration results in a broad and ultrafast spectral response spanning from UV-to-NIR, operating in a self-powered mode [10]. The ability of photodetectors to achieve broadband spectral response while reducing costs and minimizing spatial requirements is highly advantageous for advanced applications, including healthcare, military operations, weather monitoring, and imaging applications.

However, single-layer graphene-based photodetectors exhibit significant limitations, including a high level of dark current density, the lowest photocurrent responsivity, and compromised photosensitivity. These limitations stem from graphene's zero-bandgap and its inherently low optical absorption of 2.3%, significantly hindering its applicability in optical communication systems [18, 26, 27]. Dark current density can be mitigated by employing multiple graphene layers and introducing a bandgap through chemical doping techniques [18 - 20]. Furthermore, chemical methods [10, 18 - 20, 24] can induce p-type doping in graphene layers, shifting the Fermi level closer to the valence band. This shift enhances the barrier height, leading to the formation of a wider depletion region and a stronger internal electric field at the junction. The presence of such a high electric field facilitates efficient separation and drift of charge carriers, leading to high device efficiency [18, 19, 28]. Given these advantages, few-layer graphene (FLG), exhibiting bandgap opening, high optical transmittance, and high carrier mobility [17], is employed in this work. However, very few reported studies have demonstrated the use of FLG and other semiconductor-based dual-heterojunctions as broadband photodetectors.

The integration of graphene, characterized by its high carrier mobility, with ZnO, a material known for its strong ultraviolet light absorption, creates a synergistic effect that significantly enhances the photodetector's performance characteristics. Additionally, using Si as the substrate ensures compatibility with current semiconductor technologies, allowing for large-scale and cost-effective manufacturing of these devices. Furthermore, FLG serves multiple purposes- it acts as a special electrical contact called a Schottky contact, a transparent conductor, or even a template for growth.

The electric field profile at the FLG-ZnO interface efficiently splits light-generated electron-hole pairs, enabling the device to function without external power. By using heavily p-doped FLG (p^+-FLG), the device achieves better charge separation thanks to tunable Fermi levels, while also minimizing light reflection. The transparency of FLG across different wavelengths allows light to reach both the ZnO layer and the Si substrate effectively.

This study aims to explore the optoelectronic characteristics of a broadband photodetector utilizing the p^+-FLG/n-ZnO/n-Si dual-heterojunction. The study is conducted at room temperature using the Silvaco Atlas TCAD software. This research evaluates the photodetector's performance metrics, including bandgap energy diagram, built-in electric field profile, current density-voltage characteristics, resistance area-voltage characteristics, photoswitching characteristics, frequency response curve, spectral response, quantum efficiency, photocurrent responsivity, detectivity, and noise equivalent power across the UV-to-NIR spectrum region.

Moreover, the design and optimization of high-performance photodetectors pose considerable challenges, as they necessitate extensive experimental exploration of different device parameters, such as materials, geometry, and doping profiles. Each parameter needs to be carefully validated and optimized, making the design process complex. The complex interplay between a photodetector's structural elements, material properties, and optoelectronic characteristics necessitates the use of sophisticated modeling techniques for accurate performance prediction. These characteristics include a wide range of the above-mentioned electrical and optical parameters.

Machine learning (ML), a subbranch of artificial intelligence, offers significant promise for effectively overcoming these challenges. Through the analysis of extensive data sets, ML algorithms can iteratively refine models, enabling more precise predictions and enhanced decision-making capabilities. Additionally, ML helps in gaining deeper insights into the underlying patterns and behaviors exhibited by optoelectronic devices. Regression analysis, a statistical technique commonly used in ML, plays a pivotal role in understanding the relationship between various variables. For instance, it helps determine how variations in independent variables, including applied voltage and incident light wavelength, affect dependent variables like photodetector current density and spectral response. By analyzing data and quantifying dependencies, regression models provide valuable insights into the factors that shape photodetector performance characteristics.

Due to their remarkable ability to extract precise predictions from intricate datasets associated with devices and materials, ML regression models have become indispensable tools in the fields of nanomaterials and materials science [29]. These ML techniques effectively enable the extraction of valuable insights and the uncovering of the underlying relationships that govern the optoelectronic behavior and performance of photodetectors. As a result, ML demonstrates its suitability as a powerful tool for the exploration and refinement of photodetector designs. It streamlines the prediction of device properties that would otherwise require computationally intensive theoretical calculations. Machine learning methods help speed up photodetector development by streamlining design and optimization processes, allowing researchers to create devices with specific performance requirements more quickly. This chapter aims to contribute to the existing research by conducting a comprehensive comparative analysis of well-known regression models applied to the photodetector design, thereby filling a gap in current knowledge. Our novel contribution focuses on identifying the most effective method for design optimization and enabling predictive modeling of this photodetector, particularly for imaging applications. To achieve this objective, ML models are employed for the simulated data. The simulated dataset is initially divided into distinct training and testing sets. Subsequently, ML regression models are used to accurately forecast the key performance metrics of photodetection.

To comprehensively assess and compare the performance of various ML models, a set of evaluation metrics is utilized, including Mean Squared Error (MSE), Mean Absolute Error (MAE), Root Mean Squared Error (RMSE), and the coefficient of determination (R^2). These performance metrics serve as distinct indicators of a regression model's predictive power for the target variable when considering the input features. Model performance is evaluated by quantifying the deviation between empirical photodetector performance data and the model's predictions, utilizing test sets comprising 40% to 70% of the total dataset.

This research undertakes a comparative assessment of several regression models, namely decision trees (DT), k-nearest neighbors (KNN), random forests (RF), and extra trees (ET). These models are selected for their capacity to effectively model both linear and non-linear dependencies in the data, enabling accurate predictions of photodetector characteristics from the given input parameters.

By exploiting the strengths of these diverse regression models and evaluating their performance using a comprehensive set of metrics, this research endeavors to pinpoint the most efficient strategy for optimizing the design and maximizing the predictive modeling capabilities of the proposed broadband photodetector, ultimately enhancing its suitability for high-performance imaging applications.

This chapter aims to contribute value to designing and developing next-generation high-performance broadband photodetectors for high-frequency operations by systematically analyzing the proposed dual-heterojunction and utilizing machine-learning insights.

Proposed Photodetector Structure

This chapter aims to investigate and propose a broadband photodetector device design based on a dual-heterojunction structure. The proposed structure consists of a junction formed between p^+-FLG and n-ZnO, which is then integrated on an n-Si substrate, as depicted in the schematic of Fig. (**1**). The proposed p^+-FLG/n-ZnO/n-Si dual-heterojunction photodetector has a top-illuminated configuration with an illumination intensity (P_{in}) of 1 W/cm_2. The p^+-FLG layer is considered to have a thickness of 2 nm, equivalent to approximately 6 layers of graphene, with an energy bandgap (E_g) of 0.25 eV, as supported by previous studies [10, 17 - 20, 23, 24]. The n-ZnO layer and n-Si substrate thicknesses are fixed at 500 nm and 5 μm, respectively. Aluminum serves as the material employed for forming the ohmic electrical contacts, functioning as both the anode and cathode in the device structure. The photodetector area is set to 40 μm_2 during the simulation.

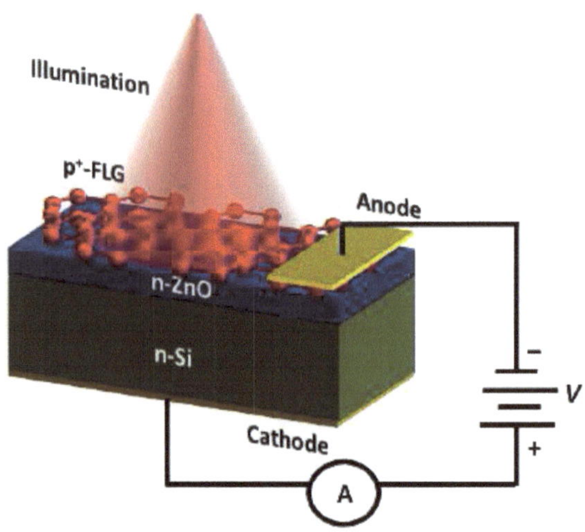

Fig (1). The schematic of p^+-FLG/n-ZnO/n-Si based dual-heterojunction photodetector.

To analyze the spectral response of the photodetector, the BEAM statement, which encompasses the entirety of the illuminated device area, is utilized. Uniform doping is applied to all the regions. The introduction of p-type doping within the FLG can be realized through the implementation of chemical processes that effectively shift the Fermi level of the FLG closer to the valence band energy level [18]. The increased barrier height leads to the generation of a large electric field at the p^+-FLG/n-ZnO interface. This electric field facilitates efficient separation and transport of charge carriers, consequently enhancing the overall photodetector efficiency [17].

In this work, the p⁺-FLG region is doped with a concentration of 2×1022 cm–3 [17, 18], the n-ZnO region with 1016 cm–3 [10], and the n-Si substrate with 1015 cm–3 [10]. The other simulation parameters, including intrinsic carrier concentration (ni), electron affinity (χ), relative permittivity (εr), effective density of states for holes in the valence band (NVB), effective density of states for electrons in the conduction band (NCB), electron lifetime (τn), hole lifetime (τp), electron mobility (μn), and hole mobility (μp) for the proposed broadband photodetector, have been adopted from a previous study [10, 13] and are presented in Table **1**. The optoelectronic characteristics are evaluated at room temperature. This evaluation involves implementing the continuity equation, carrier transport diffusion equations, and the Poisson equation, utilizing Boltzmann's transport model with appropriate boundary conditions. To accurately model device behavior, conventional mechanisms, including optical, Auger, and Shockley-Read-Hall models, are incorporated into the simulation framework.

Table 1. Simulation parameter values utilized for the analysis of the photodetector at a temperature of 300 K.

Parameters	p^+-FLG	n-ZnO	n-Si
E_g (eV)	0.25	3.37	1.12
χ (eV)	4.8	4.5	4.05
ε_r	3.3	8.5	11.9
n_i (cm^{-3})	3.18×10^{15}	8.26×10^{-10}	1.42×10^{10}
N_{CB} (cm^{-3})	4×10^{17}	4×10^{18}	6×10^{19}
N_{VB} (cm^{-3})	4×10^{17}	7×10^{19}	2.2×10^{19}
τ_n (s)	5×10^{-12}	10^{-9}	10^{-3}
τ_p (s)	5×10^{-12}	10^{-9}	10^{-3}
μ_n (cm^2/Vs)	40000	60	1000
μ_p (cm^2/Vs)	400	10	500

SILVACO ATLAS SIMULATION RESULTS AND DISCUSSIONS

In this section, the electrical and optical characteristics of a UV-to-NIR dual-heterojunction broadband photodetector, consisting of an optimized configuration of p^+-FLG/n-ZnO/n-Si, are evaluated. When exposed to the illuminations, the graphene layers, ZnO, and Si absorb incident photons at different wavelengths, generating charge carriers within their respective band gaps.

The internal electric field at the heterojunctions facilitates the separation of these charge carriers. Furthermore, the applied external electric field aligns with the overall electric field, causing the photo-generated carriers to drift toward the electrodes. This motion of photo-excited carriers results in a net photocurrent that passes through the external circuit.

Fig. (**2**) illustrates the simulated energy bandgap profile of the proposed broadband photodetector at 0 V, *i.e.*, in the self-powered mode of operation. Fig. (**3**) demonstrates the electric field (Efield) profile of the photodetector at 0, –1.0, –2.0, and –5.0 V. It is observed that there is a positive correlation between the bias voltage and the electric field strength within the photodetector device. The E-field in the proposed broadband photodetector is calculated to be 17.93, 37.83, 57.70, and 116.85 kV/cm at 0, –1.0, –2.0, and –5.0 V, respectively. This enhanced electric field expedites the transport of photo-induced charge carriers toward their respective electrodes, enabling the efficient collection of the generated charges.

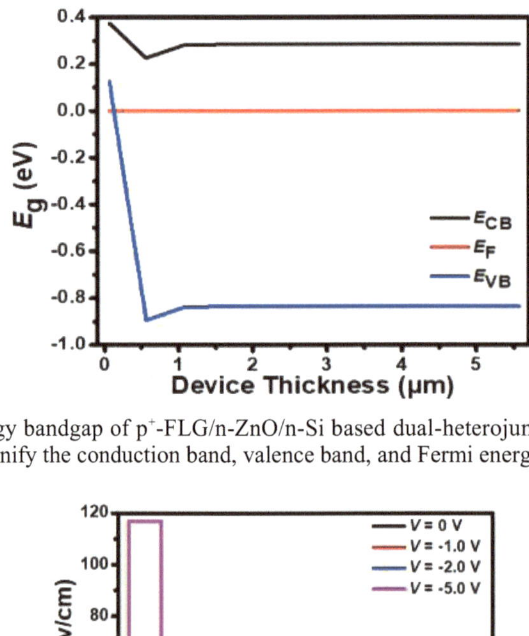

Fig (2). Simulated energy bandgap of p⁺-FLG/n-ZnO/n-Si based dual-heterojunction photodetector at 0 V. Here, E_{CB}, E_{VB}, and E_F signify the conduction band, valence band, and Fermi energy levels, respectively.

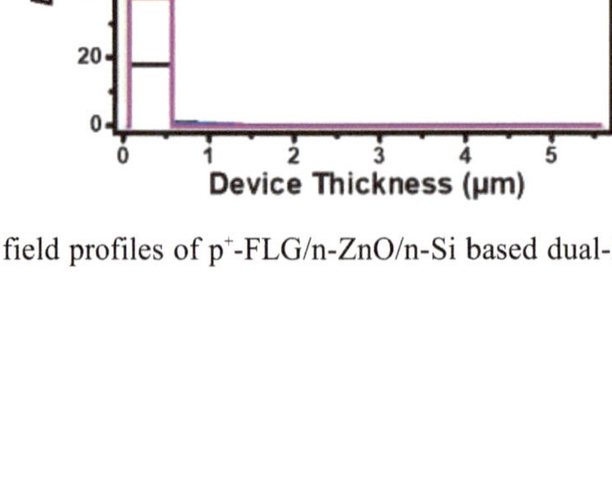

Fig. (3). Simulated electric field profiles of p⁺-FLG/n-ZnO/n-Si based dual-heterojunction photodetector at different bias voltages.

The standard dark current density (Jdark) versus voltage (V) characteristics of the proposed broadband photodetector, demonstrated on both linear and logarithmic scales, is demonstrated in Fig. (**4**). The photodetector exhibits a Jdark of 1.8×10¯19 A/cm^2 at 0 V and 2.7×10¯15 A/cm^2 at –1.0 V. These measured Jdark values for the proposed broadband photodetector outperform those of previously reported photodetectors [8, 16].

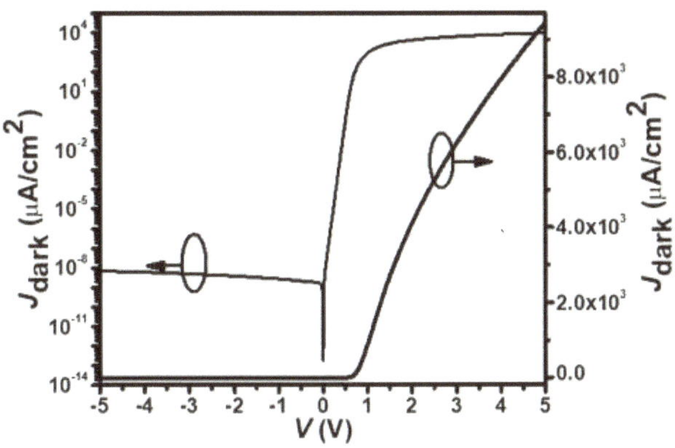

Fig. (4). The simulated J_{dark}-V characteristics of p$^+$-FLG/n-ZnO/n-Si based dual-heterojunction photodetector in linear and logarithmic scale.

Fig. (**5**) presents the resistance-area (RA)-V characteristics in both dark and illuminated conditions using a logarithmic scale. When exposed to illumination, the RA in the photodetector device decreases, facilitating the generation of charge carriers induced by the absorption of photons. The heterojunction formed by p$^+$-FLG, n-ZnO, and n-Si demonstrates a zero-bias RA of 1.6×107 MΩcm^2 in the dark and 117.6 MΩcm^2 under illumination. A high zero-bias resistance-area product (RA) is desirable as it directly correlates with a high specific detectivity for the photodetector.

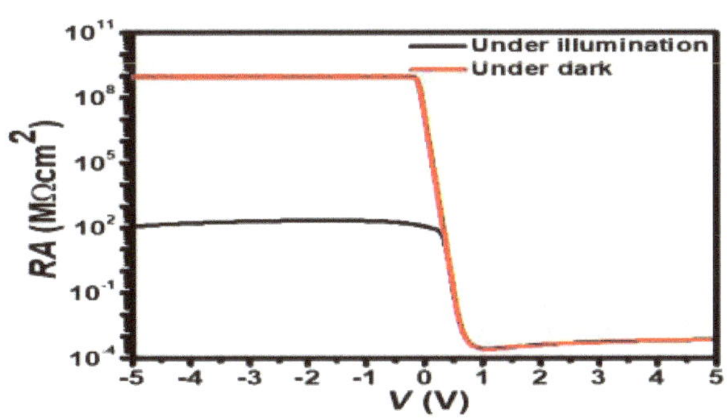

Fig. (5). The simulated *RA-V* characteristics of p$^+$-FLG/n-ZnO/n-Si based dual-heterojunction photodetector under dark and illuminated conditions in logarithmic scale.

A three-dimensional (3D) plot representing the photocurrent density (J_{light}) concerning the applied bias voltage and wavelength (λ) for the proposed photodetector at room temperature, with an illumination intensity of 1 W/cm^2 is shown in Fig. (**6**). The J_{light}-λ characteristic curve reveals an enhanced response within the 280 to 900 nm range at a bias of –1.0 V, as depicted. The proposed broadband photodetector exhibits a total J_{light} that is a combined response from both the ZnO and Si components. This confirms the broadband photodetection capability of the photodetector in the UV-to-NIR region. The spectral response curve (Jlight-λ) depicted in Fig. (**6**) exhibits a maximum before reaching the cut-off wavelength. This behavior arises from the fundamental limitation imposed by the bandgap of the photosensitive material, which restricts the absorption of incident photon energy beyond a specific wavelength. When exposed to UV light, the primary light absorption occurs in the ZnO material, generating photo-induced carriers. For the proposed photodetector, when illuminated with light in the visible to IR wavelength range, ZnO functions as a transparent material owing to its wide-bandgap. Consequently, all photons in the visible to NIR range are absorbed in the n-Si region, creating electron-hole pairs. This absorption and subsequent charge carrier generation process enables the photodetector to produce photocurrent under reverse bias conditions.

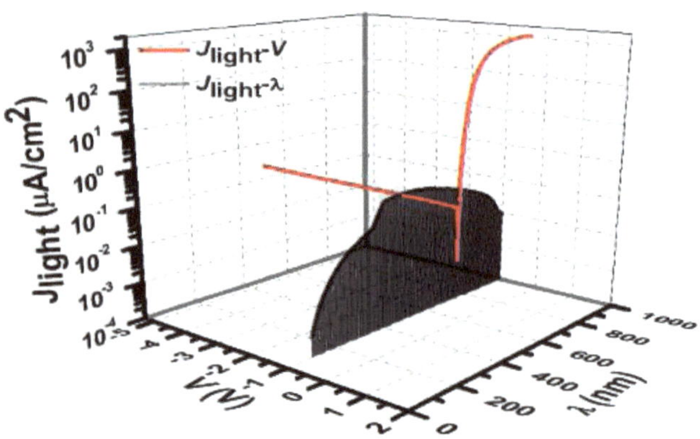

Fig. (6). The 3D plot of J_{light}-V and J_{light}-λ of p$^+$-FLG/n-ZnO/n-Si based dual-heterojunction photodetector in logarithmic scale.

As evident from Figs. (**4** and **6**), the proposed broadband photodetector demonstrates a lower current density in reverse bias than in forward bias, indicating the creation of a junction barrier. The Jlight value for the proposed broadband photodetector is estimated to be 0.26 µA/cm2 at 0 and –1.0 V under 470 nm illumination wavelength. The photocurrent density to dark current density ratio (Jlight/Jdark), also called the on-to-off ratio of 1.43×1012 and 9.77×107, is obtained at 0 and –1.0 V, respectively.

Fig. (**7**) shows the impact of the incident wavelength on the performance of the photodetector by presenting the Jlight-V characteristics at different wavelengths. The Jlight-V curves exhibit an increasing trend as the wavelength increases from 350 nm to 470 nm. However, beyond 470 nm, the Jlight-V curves show a decreasing behavior with further increases in wavelength. This behavior can be ascribed to the cumulative photoresponse originating from both the ZnO and Si components in the proposed photodetector, which results in a maximum point at 470 nm wavelength. The observed wavelength-dependent photoresponse is similar to that of earlier reported ZnO/Si-based photodetector [30].

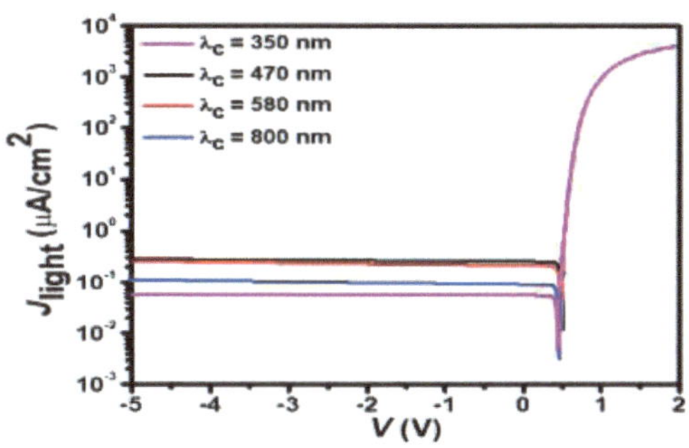

Fig. (7). The J_{light}-V characteristics of p$^+$-FLG/n-ZnO/n-Si based dual-heterojunction photodetector in logarithmic scale at different wavelengths under illumination intensity of 1 W/cm^2.

The normalized frequency response of the proposed broadband photodetector is presented in Fig. (**8**). When operating at –1.0 V bias and under illumination at a wavelength of 470 nm, the device exhibits a 3-dB cut-off frequency [18] of 7.36 THz. This high cut-off frequency renders the proposed photodetector suitable for applications in the terahertz frequency range.

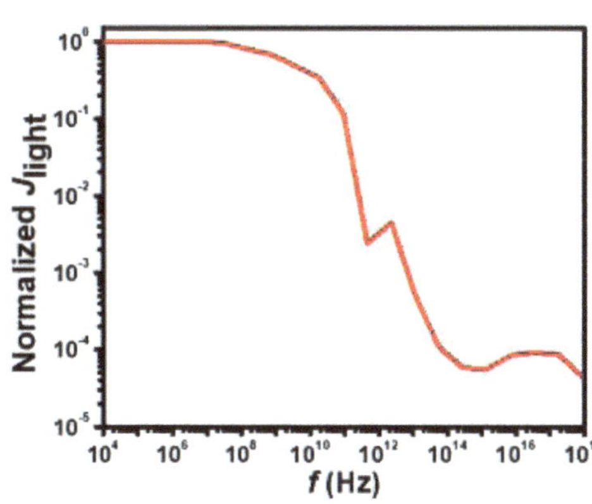

Fig. (8). Normalized frequency response curve of p$^+$-FLG/n-ZnO/n-Si based dual-heterojunction photodetector.

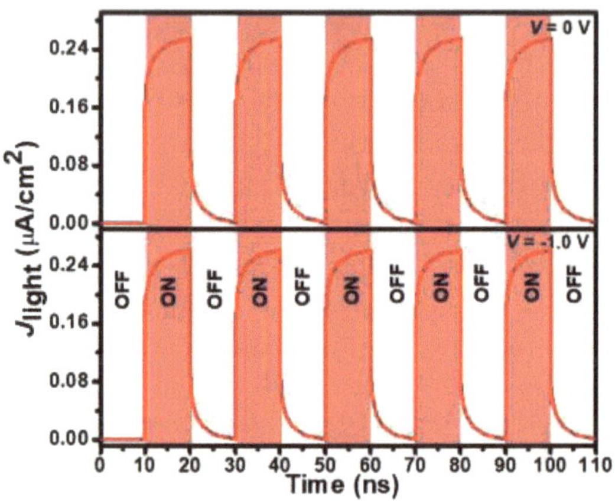

Fig. (9). J_{light}-Time characteristics of p$^+$-FLG/n-ZnO/n-Si based dual-heterojunction photodetector at 0 and −1.0 V.

The normalized frequency response of the proposed broadband photodetector is presented in Fig. (**8**). When operating at a bias voltage of −1.0 V and under illumination at a wavelength of 470 nm, the device exhibits a 3-dB cut-off frequency [18] of 7.36 THz. This high cut-off frequency renders the proposed photodetector suitable for applications in the terahertz frequency range.

Fig. (**9**) illustrates the photoswitching behavior, the relationship between the J_{light} and time, for the proposed broadband photodetector operating at 0 V and −1.0 V under illumination with a wavelength of 470 nm. The proposed photodetector exhibits a persistent photoswitching behavior characterized by a gradual exponential increase and decrease in current density when transitioning between illuminated and dark conditions, respectively. The response time, which encompasses both the rise and fall times, for the proposed broadband photodetector is less than 1 ns. Specifically, the photodetector exhibits rapid photoswitching with a rise (fall) time of 00.48 (0.23) ns and 47 (0.88) ns at 0 and −1.0 V, respectively. This response time outperforms the values reported in previous studies [8, 16, 31], indicating superior performance. The enhanced photoswitching response, characterized by improved rise and fall times, enhances the imaging capability of the photodetector. Fig. (**10**) shows the J_{light}-Time characteristics of the proposed photodetector under different illumination intensities. Fig. (**11**) demonstrates a linear relationship between the J_{light} and the incident power intensity P_{in}, suggesting negligible recombination losses in the proposed photodetector [18].

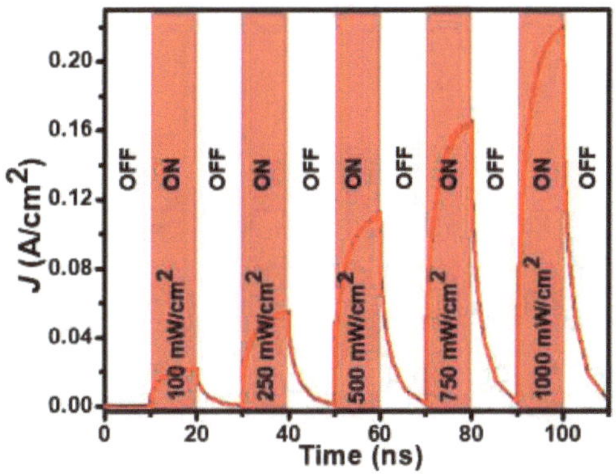

Fig. (10). J_{light}-Time characteristics of p^+-FLG/n-ZnO/n-Si based dual-heterojunction photodetector at different illumination intensities.

Figs. (**12-15**) depict the variation of optical characteristic parameters, such as internal quantum efficiency (QE_{int}), external quantum efficiency (QE_{ext}), and external photocurrent responsivity Riext, specific detectivity (D*), and noise equivalent power (*NEP*) [18], with wavelength at room temperature, under –1.0 V bias with an illumination intensity of 1 W/cm^2, respectively. In this chapter, *NEP* is estimated at 1 Hz. The maximum quantum efficiency is achieved within the wavelength region where the incoming photon energy surpasses the bandgap of the semiconductor material. Figs. (**12-14**) clearly show that the response exhibits a decreasing trend with increasing wavelength after reaching a maximum value at the cut-off wavelength of 470 nm. The proposed dual-heterojunction photodetector demonstrates a QE_{int} (QE_{ext}) of 21 (20) % at 350 nm (UV), 84.25 (69.3) % at 470 nm (visible), and 44.52 (14.72) % at 800 nm (NIR) under –1.0 V bias.

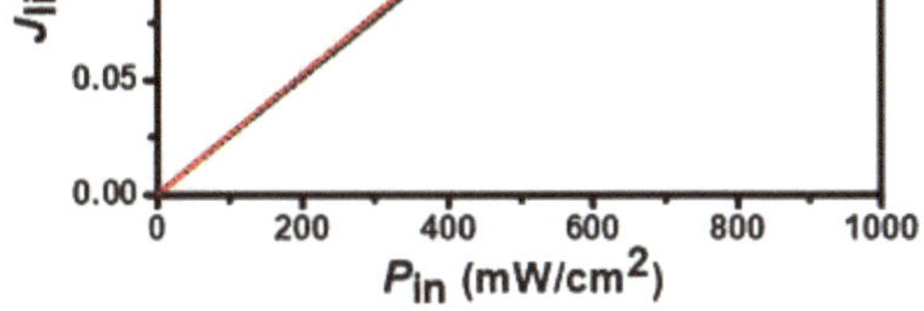

Fig. (11). J_{light}-P_{in} characteristics of p^+-FLG/n-ZnO/n-Si based dual-heterojunction photodetector at 0 and –1. 0 V under 470 nm illuminating wavelength.

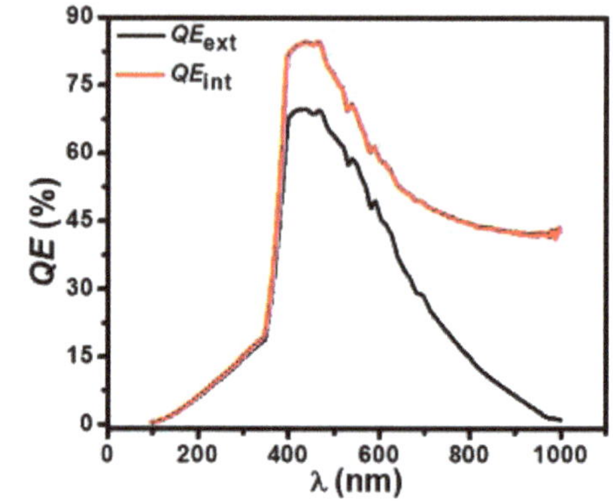

Fig. (12). The *QE*-λ response of p⁺-FLG/n-ZnO/n-Si based dual-heterojunction photodetector at –1.0 V with 1 W/cm^2 illumination intensity.

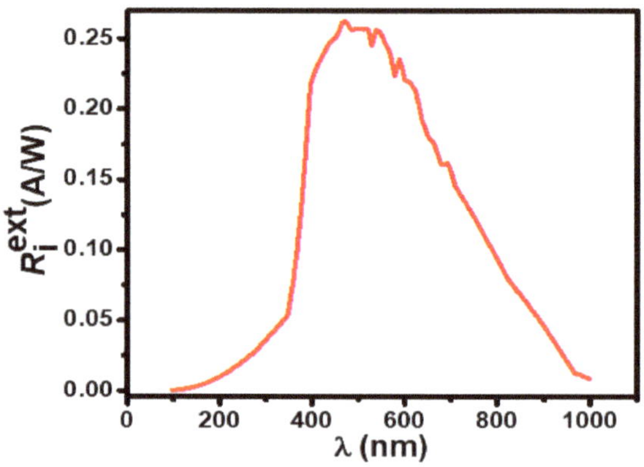

Fig. (13). The Riext-λ response of p⁺-FLG/n-ZnO/n-Si based dual-heterojunction photodetector at –1.0 V with 1 W/cm^2 illumination intensity.

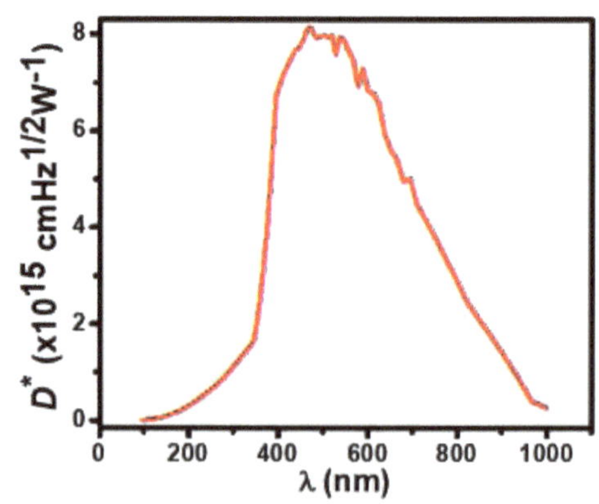

Fig. (14). The *D**-λ response of p⁺-FLG/n-ZnO/n-Si based dual-heterojunction photodetector at –1.0 V with 1 W/cm^2 illumination intensity.

The photodetector achieves a peak Riext of 0.26 A/W, a maximum *D** of 8.12×10^{15} cmHz$^{1/2}$/W, and the lowest *NEP* of 7.79×10^{-20} W at 470 nm. The decline in *QE*, Riext, and *D** with increasing wavelength can be attributed to a rise in the rate of charge carrier recombination processes. The optical characteristic parameter values obtained for the proposed photodetector outperform the values reported in previous studies [8, 16, 32 - 35].

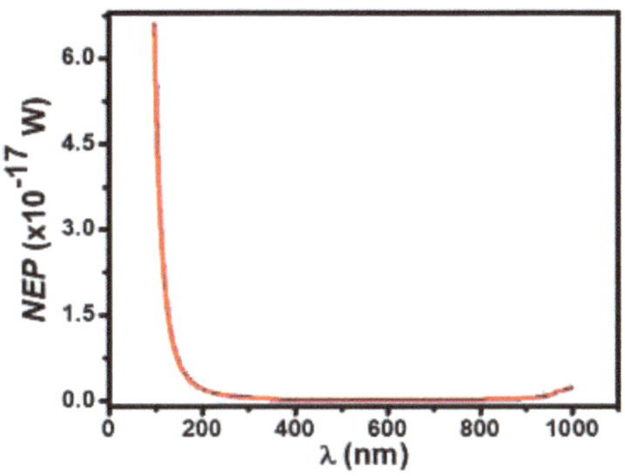

Fig. (15). The *NEP*-λ response of p$^+$-FLG/n-ZnO/n-Si based dual-heterojunction photodetector at –1.0 V with 1 W/cm^2 illumination intensity.

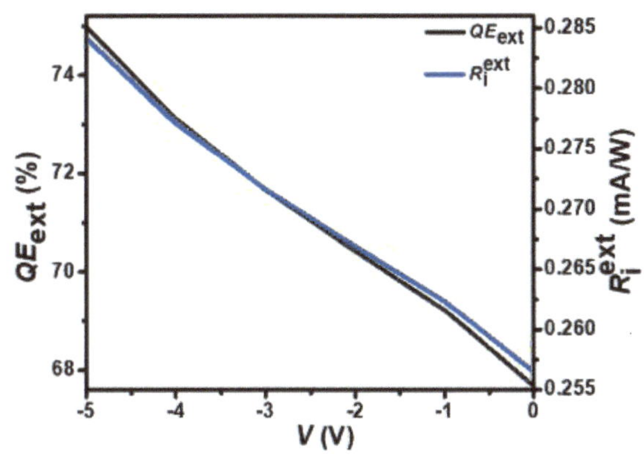

Fig. (16). The variation of QE_{ext} and Riext with the applied reverse bias of p$^+$-FLG/n-ZnO/n-Si based dual-heterojunction photodetector with 1 W/cm^2 illumination intensity.

Fig. (**16**) depicts the QE_{ext} and Riext of the proposed photodetector as a function of the reverse bias voltage, whereas Fig. (**17**) shows the variation of *D** and *NEP* with respect to the reverse bias at 470 nm wavelength. The enhancement in the optical performance of the photodetector with increasing reverse bias voltage can be attributed to several factors. The contribution of photogenerated charge carriers increases, leading to a higher photocurrent. Additionally, the probability of exciton separation is enhanced, facilitating more efficient charge carrier generation. Furthermore, the drift velocity of the charge carriers, which is bias-dependent, also increases, facilitating more efficient carrier transport and collection in the external circuit, thus contributing to the overall photocurrent. As the reverse bias voltage varies from 0 to –5.0 V, the QE_{ext} increases from 67.67 to 74.98%, Riext rises from 0.26 to 0.28 A/W, D^* improves from 7.94×10^{15} to 8.79×10^{15} cmHz$^{1/2}$/W, and *NEP* decreases from 7.97×10^{-20} to 7.19×10^{-20} W.

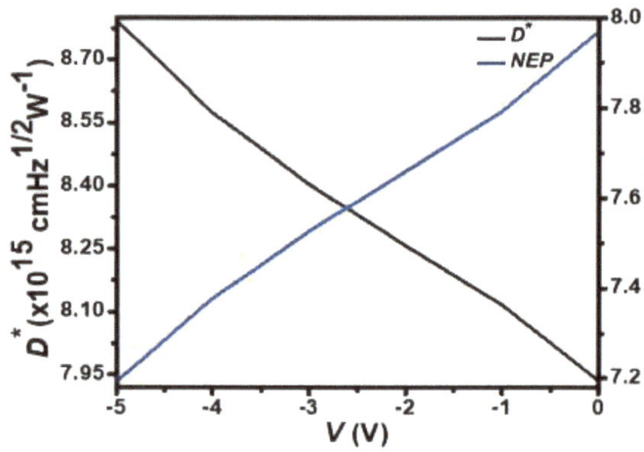

Fig. (17). The variation of D^* and *NEP* with the applied reverse bias of p$^+$-FLG/n-ZnO/n-Si based dual-heterojunction photodetector with 1 W/cm^2 illumination intensity.

Fig. (**18**) depicts the noise current components with respect to the applied bias voltage, while Fig. (**19**) illustrates the corresponding behavior of the signal-to-noise ratio (*SNR*) [18, 24] as a function of the applied bias voltage. The total noise current (i_n) comprises the Johnson noise current (i_j) and the shot noise current (i_s) [18]. As depicted in Fig. (**18**), the shot noise exhibits a noticeable increase with the applied bias voltage, primarily due to the high dark current density, which emerges as the dominant source of noise.

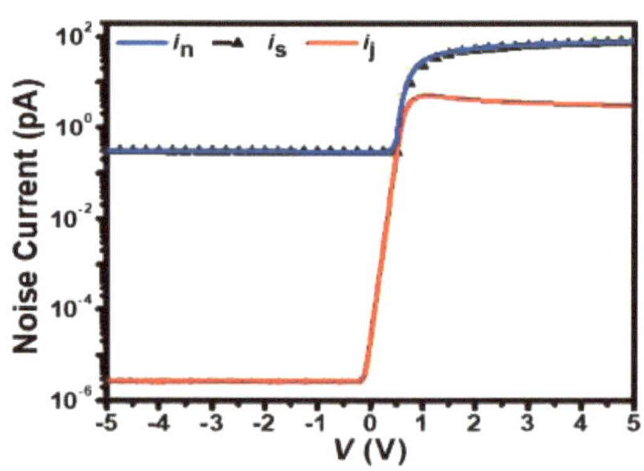

Fig. (18). The variation of noise current with the applied reverse bias of p$^+$-FLG/n-ZnO/n-Si based dual-heterojunction photodetector.

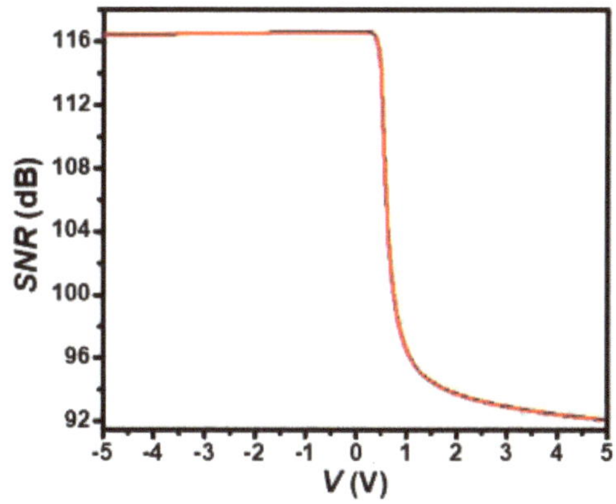

Fig. (19). The variation of *SNR* with the applied reverse bias of p^+-FLG/n-ZnO/n-Si based dual-heterojunction photodetector.

In contrast, the impact of Johnson's noise on the overall noise is minimal compared to the shot noise, indicating that the presence of a heterojunction effectively filters out the undesired thermal generation-recombination current component. The net noise current is found to be 0.29 pA. The *SNR* for the proposed photodetector is estimated to be 116.41 dB. This exceptional photodetection performance is attributed to the enhanced light absorption by the ZnO and Si layers, as well as the substantial electric field at the heterojunction. This electric field facilitates the efficient movement of photogenerated carriers toward the electrodes, enabling the efficient collection of generated charges. These results suggest that the proposed dual-heterojunction photodetector has significant potential for various cost-effective, low-noise, and high-frequency applications across the UV-to-NIR spectral region.

APPLICATIONS OF MACHINE LEARNING

By employing ML techniques, regression models emerge as invaluable tools for simulating complex systems. These models offer substantial time and resource savings while providing accurate forecasts of critical parameters and efficiently handling missing data, thereby streamlining the analysis process. The development of such models typically entails simulating the design under investigation and generating an extensive dataset that encapsulates the relevant features and target variables. This comprehensive data compilation is subsequently employed to train the regression model, enabling it to learn the inherent patterns and correlations between the input features and the desired outputs. Once trained, the regression model efficiently predicts values for previously unseen data points, thereby avoiding the need for computationally expensive simulations. In this chapter, supervised ML techniques are employed due to their suitability for addressing well-defined objective problems, rather than employing unsupervised or reinforcement ML approaches [38, 39]. The specific objective of this research aligns perfectly with the nature of supervised learning, making it the most appropriate choice among ML paradigms.

Python has emerged as a popular programming language for implementing regression models due to its vast ecosystem of ML libraries and tools. These libraries provide various regression algorithms, enabling researchers and practitioners to select the most suitable technique for their problem domain. Utilizing simulated data and ML regression models within the Python framework revolutionizes the analysis process for researchers. This approach enables more efficient exploration of complex systems and facilitates rapid iteration in design optimization.

Simple Machine-Learning Regression Models

The comprehensive dataset generated serves as training data for the regression model, enabling it to uncover the underlying patterns and relationships between input features and target outputs. After the training phase, the regression model demonstrates the capacity to accurately predict values for novel, unseen data points with efficiency.

DT is a powerful tool in data analysis, enabling predictive modeling by recursively splitting data into branches based on specific conditions until a final prediction is reached [40]. As a supervised machine learning technique, DTs mimic human decision-making, making them intuitive and easy to understand. As a straightforward and accessible method for ML, the decision tree can be employed for both classification and regression tasks. By iteratively dividing the data into smaller subsets, DTs aim to achieve homogeneity within each subset. This makes DT a viable option for regression problems with limited data, although careful tuning is crucial to prevent overfitting.

The KNN algorithm is a non-parametric, supervised ML technique that finds widespread application in both classification and regression problems, owing to its simplicity and flexibility. In contrast to parametric methods, which make assumptions about the underlying data distribution, KNN does not rely on such assumptions, rendering it robust and adaptable. For classification problems, KNN classifies a new instance by determining the most frequent class among its nearest neighbors in the feature space. Conversely, in regression problems, KNN predicts the outcome by averaging the values of the nearest neighboring instances. Despite its conceptual simplicity, the performance of KNN can be enhanced through optimization of the k value, differential weighting of neighbors, and the use of advanced distance metrics. These modifications enable KNN to be effectively applied across a wide range of domains, where accurate and reliable predictions are crucial [41].

Bagging-Based Ensemble Machine-Learning Regression Models

Ensemble methods like Random Forest (RF) and Extra Trees (ET), which are based on bagging, improve performance and robustness by combining predictions from multiple base models. Bagging employs bootstrapping to train each base model on a different, randomly sampled subset of the training data. This technique reduces overfitting and variance by averaging the predictions, leading to more stable and accurate results. Bagging is particularly beneficial for high-variance models like Decision Trees, effectively mitigating their tendency to overfit the training data.

RF is a powerful ML technique that delivers accurate predictions while mitigating overfitting. It constructs multiple decision trees using bootstrapping, where each tree is trained on a random subset of predictors, reducing correlation and increasing randomness. RF further enhances this by considering only a subset of predictors at each node for splitting, rather than evaluating all variables [42].

ET, an extension of RF, builds a collection of fully grown decision trees. ET distinguishes itself by utilizing all potential splitting points and training each tree on the entire dataset, minimizing bias. Moreover, ET randomly selects splitting points at each node, improving accuracy and reducing overfitting [42].

The performance of these regression models can be assessed using various metrics, including MSE, RMSE, MAE, and R^2. These metrics provide crucial insights into the model's generalization capability, how well it fits the data, and its accuracy in predicting the behavior of the proposed photodetector. The following equations are employed for the calculation of these performance indices [42 - 44]:

$$\boldsymbol{MSE} = \frac{1}{N}\sum_{i=1}^{N}\left(\text{Actual\_Value}_i - \text{Predicted\_Value}_i\right)^2 \quad \textbf{(1)}$$

$$\boldsymbol{RMSE} = \sqrt{\frac{1}{N}\sum_{i=1}^{N}\left(\text{Actual\_Value}_i - \text{Predicted\_Value}_i\right)^2} \quad \textbf{(2)}$$

$$\boldsymbol{MAE} = \frac{1}{N}\sum_{i=1}^{N}\left(\frac{\text{Actual\_Value}_i - \text{Predicted\_Value}_i}{\text{Actual\_Value}_i}\right) \quad \textbf{(3)}$$

$$\boldsymbol{R^2} = 1 - \frac{\sum_{i=1}^{N}\left(\text{Predicted\_Value}_i - \text{Actual\_Value}_i\right)^2}{\sum_{i=1}^{N}\left(\text{Actual\_Value}_i - \text{Average\_Target\_Value}_i\right)^2} \quad \textbf{(4)}$$

where N represents the total number of data samples employed for validating the performance of the regression model.

MSE and RMSE measure the average error magnitude, with higher values indicating poorer performance. MAE calculates the average absolute error, providing an easily interpretable metric for assessing prediction accuracy. Conversely, R^2 assesses how well the model explains the variability in the target variable. The values closer to 1 indicate a better model fit.

Machine-Learning Results and Discussions

The scatter plots depicted in Figs. (**20-23**) represent the correlation between the predicted values of J_{dark}, J_{light}, QE_{ext}, and Riext by DT, KNN, RF, and ET regressors, respectively, and the corresponding simulated values. Scatter plots are generated for different test set sizes, varying from 40% to 70%, which indicates the proportion of data used for testing the models' performance.

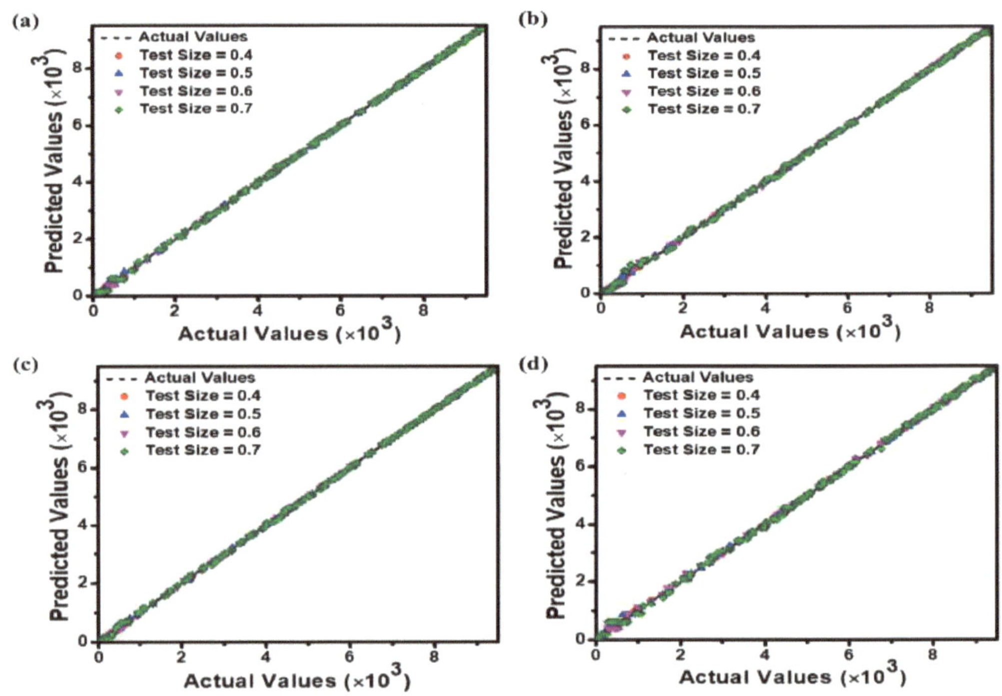

Fig. (20). Scatter plots showing the association between predicted and simulated J_{dark} values for different test set sizes using **(a)** DT, **(b)** KNN, **(c)** RF, and **(d)** ET algorithms.

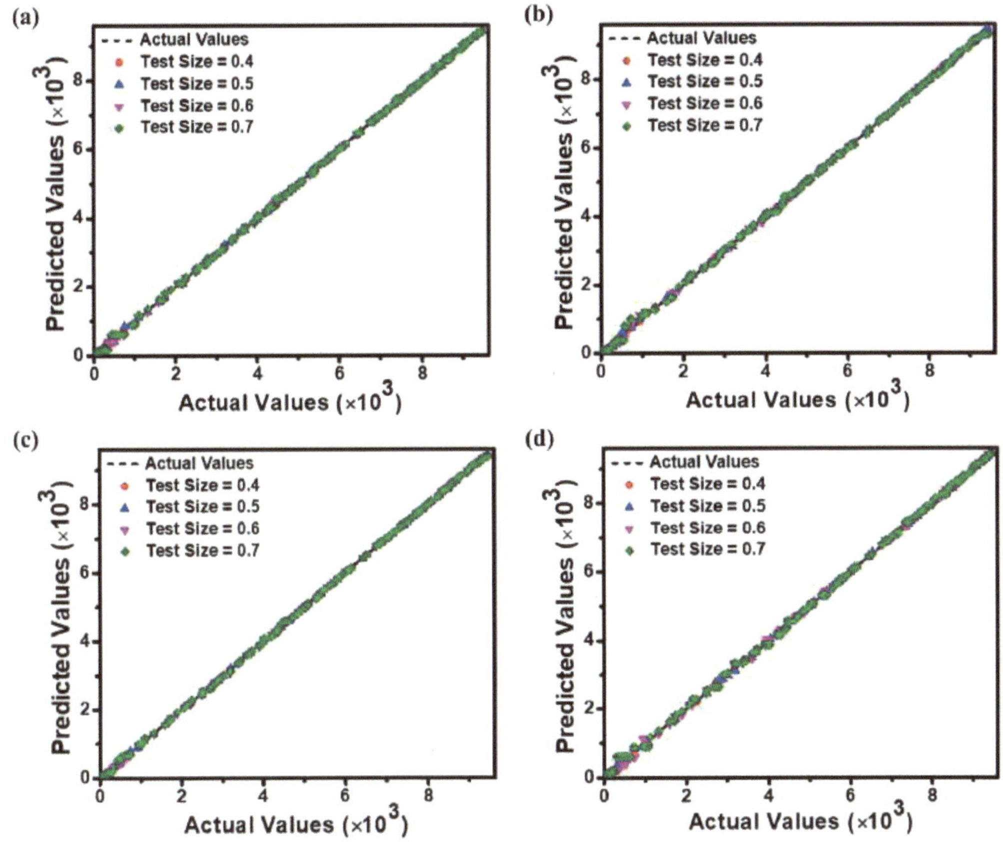

Fig. (21). Scatter plots showing the association between predicted and simulated J_{light} values for different test set sizes using **(a)** DT, **(b)** KNN, **(c)** RF, and **(d)** ET algorithms.

The evaluated models, the ET regressor, overwhelmingly exhibits the highest accuracy for estimating the proposed broadband photodetector's electrical and optical characteristic parameters. These figures illustrate the performance of ML regression models in forecasting these parameters across various test set sizes. The scatter plots provide a visual representation of the model's predictions, estimation errors, and general patterns. By incorporating multiple test set sizes, the study enables an in-depth exploration of the models' behavior across varying ratios of data used for training and evaluation.

Tables **2** to **9** present a comprehensive evaluation of the performance of ML regression models in predicting the optoelectronic characteristics of a broadband photodetector. This evaluation is conducted by assessing performance metrics that quantify the deviations between predicted and actual values. Test set sizes for this evaluation range from 40% to 70% of the entire dataset.

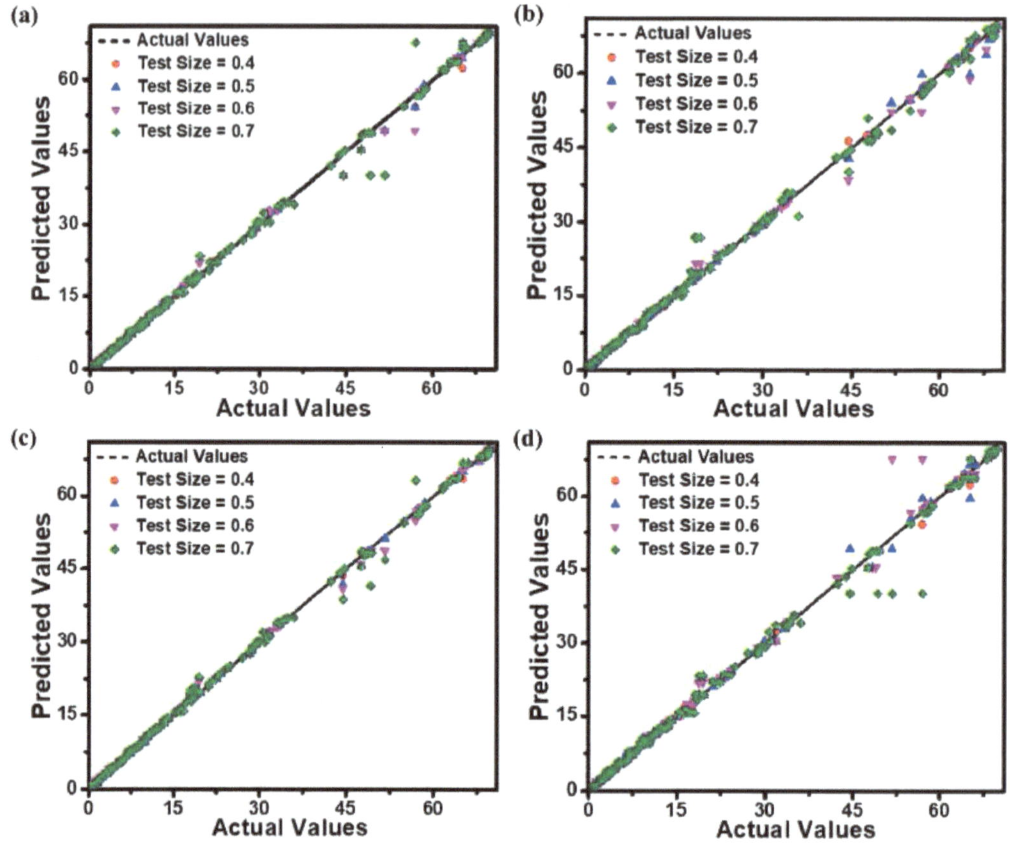

Fig. (22). Scatter plots showing the association between predicted and simulated QE_{ext} values for different test set sizes using **(a)** DT, **(b)** KNN, **(c)** RF, and **(d)** ET algorithms.

Tables **2** to **9** indicate that DT and ET regression models demonstrate superior predictive accuracy, as indicated by their lower mean squared error (MSE), root mean squared error (RMSE), and mean absolute error (MAE) values. Notably, both models attain an exceptional R^2 of 1 for all test set sizes. This near-perfect R^2 score signifies that DT and ET models effectively capture data variability, leading to excellent agreement between predicted and observed photodetector parameter values. The consistently low error rates and high R^2 values underscore the robust predictive power of these models, even when trained on smaller datasets. These findings emphasize the effectiveness of various ML models and emphasize the importance of model selection based on dataset characteristics and the desired level of predictive performance.

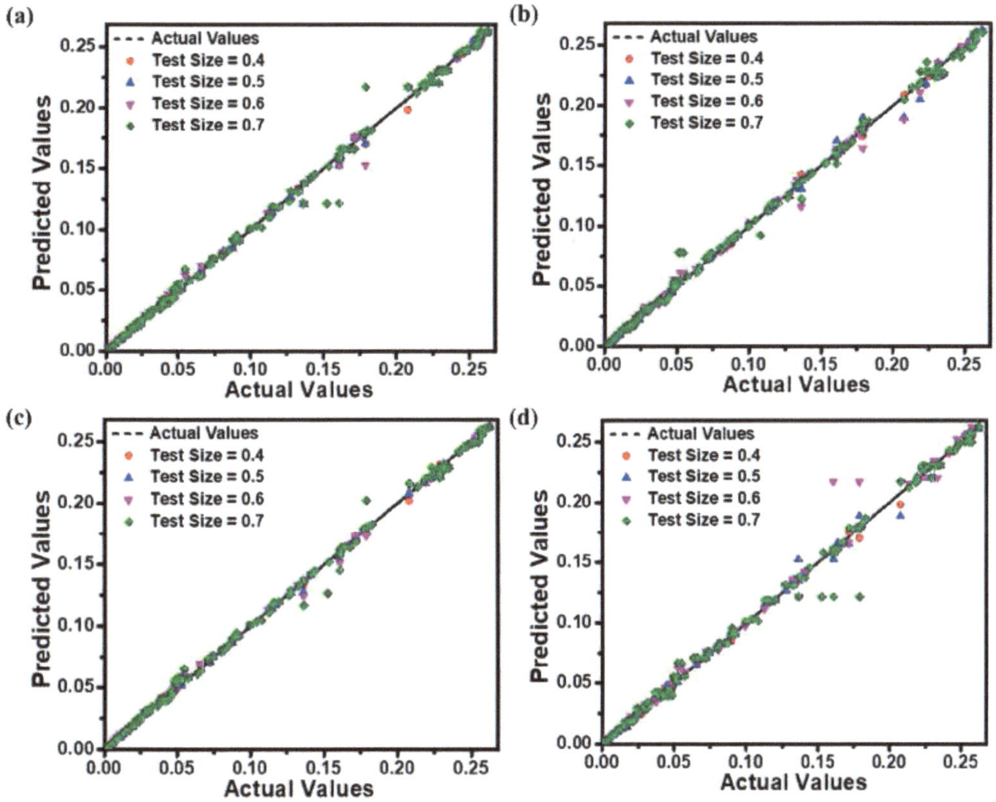

Fig. (23). Scatter plots showing the association between predicted and simulated Riext values for different test set sizes using **(a)** DT, **(b)** KNN, **(c)** RF, and **(d)** ET algorithms.

Table 2. MSE and RMSE for various ML regression models in predicting J_{dark}.

ML Regression Models	MSE				RMSE			
	Test Sizes				Test Sizes			
	0.4	0.5	0.6	0.7	0.4	0.5	0.6	0.7
DT	6.35E-17	7.52E-17	4.14E-17	2.37E-17	7.97E-09	8.67E-09	6.44E-09	4.87E-09
KNN	130.67	303.93	495.36	1011.76	11.43	17.43	22.26	31.81
RF	21.80	30.61	75.91	100.36	4.67	5.53	8.71	10.02
ET	4.12E-19	3.36E-19	3.77E-19	6.13E-19	6.42E-10	5.80E-10	6.14E-10	7.83E-10

Table 3. MAE and R^2 for various ML regression models in predicting J_{dark}.

ML Regression Models	MAE				R^2			
	Test Sizes				Test Sizes			
	0.4	0.5	0.6	0.7	0.4	0.5	0.6	0.7
DT	1.48E-09	1.62E-09	1.35E-09	1.20E-09	1	1	1	1
KNN	5.495819	7.67	10.75	15.18	0.999988	0.999972	0.999955	0.99999
RF	2.498142	2.80	3.88	4.77	0.999998	0.999997	0.999993	1
ET	3.47E-10	3.27E-10	3.33E-10	3.76E-10	1	1	1	1

Table 4. MSE and RMSE for various ML regression models in predicting J_{light}.

ML Regression Models	MSE				RMSE			
	Test Sizes				Test Sizes			
	0.4	0.5	0.6	0.7	0.4	0.5	0.6	0.7
DT	0	0	0	0	0	0	0	0
KNN	130.68	303.95	495.38	1011.81	11.43	17.43	22.26	31.81
RF	21.80	30.61	75.92	100.36	4.67	5.53	8.71	10.02
ET	3.47E-23	3.58E-23	3.63E-23	3.43E-23	5.89E-12	5.98E-12	6.03E-12	5.86E-12

Table 5. MAE and R^2 for various ML regression models in predicting J_{light}.

ML Regression Models	MAE				R^2			
	Test Sizes				Test Sizes			
	0.4	0.5	0.6	0.7	0.4	0.5	0.6	0.7
DT	0	0	0	0	1	1	1	1
KNN	5.50	7.67	10.75	15.18	0.999988	0.999972	0.999955	0.999904
RF	2.50	2.80	3.88	4.77	0.999998	0.999997	0.999993	0.99999
ET	3.10E-12	3.19E-12	3.27E-12	3.18E-12	1	1	1	1

Table 6. MSE and RMSE for various ML regression models in predicting Q_{ext}.

ML Regression Models	MSE				RMSE			
	Test Sizes				Test Sizes			
	0.4	0.5	0.6	0.7	0.4	0.5	0.6	0.7
DT	0	0	0	0	0	0	0	0
KNN	0.21	0.30	0.43	1.91	0.46	0.55	0.66	1.38
RF	0.035	0.04	0.11	0.15	0.19	0.21	0.33	0.39
ET	2.62E-27	2.47E-27	2.40E-27	1.98E-27	5.12E-14	4.97E-14	4.90E-14	4.45E-14

Table 7. MAE and R^2 for various ML regression models in predicting Q_{ext}.

ML Regression Models	MAE				R^2			
	Test Sizes				Test Sizes			
	0.4	**0.5**	**0.6**	**0.7**	**0.4**	**0.5**	**0.6**	**0.7**
DT	0	0	0	0	1	1	1	1
KNN	0.22	0.28	0.38	0.60	0.999604	0.999913	0.999179	0.996411
RF	0.09	0.11	0.13	0.16	0.999934	1	0.999787	0.999716
ET	3.44E-14	3.30E-14	3.20E-14	2.89E-14	1	0.99942	1	1

Table 8. MSE and RMSE for various ML regression models in predicting Riext.

ML Regression Models	MSE				RMSE			
	Test Sizes				Test Sizes			
	0.4	**0.5**	**0.6**	**0.7**	**0.4**	**0.5**	**0.6**	**0.7**
DT	0	0	0	0	0	0	0	0
KNN	2.90E-06	3.83E-06	5.45E-06	2.13E-05	1.70E-3	1.96E-3	2.34E-3	4.61E-3
RF	4.70E-07	5.68E-07	1.33E-06	1.84E-06	6.86E-4	7.54 E-4	1.15E-3	1.36E-3
ET	4.29E-32	3.94E-32	4.17E-32	3.92E-32	2.07E-16	1.99E-16	2.04E-16	1.98E-16

Table 9. MAE and R^2 for various ML regression models in predicting Riext.

ML Regression Models	MAE				R^2			
	Test Sizes				Test Sizes			
	0.4	**0.5**	**0.6**	**0.7**	**0.4**	**0.5**	**0.6**	**0.7**
DT	0	0	0	0	1	1	1	1
KNN	9.17E-4	1.11E-3	1.48E-3	2.27E-3	0.999667	0.999551	0.999371	0.997585
RF	3.72E-4	4.23 E-4	4.99 E-4	6.45E-4	0.999946	0.999933	0.999846	0.999791
ET	1.44E-16	1.35E-16	1.40E-16	1.34E-16	1	1	1	1

CONCLUSIONS

In conclusion, a dual-heterojunction comprising p^+-FLG/n-ZnO/n-Si as a single framework was presented for broadband (UV-to-NIR) photodetection using Silvaco Atlas TCAD software. Various electrical and optical characteristic parameters were evaluated under both zero-bias and biased modes of operation. The photodetector demonstrated an enhancement in the photocurrent density of 10^{12} and 10^7 at 0 and –1.0 V, respectively. A 3-dB cut-off frequency of 7.36 THz was obtained, making it suitable for high-frequency operations. In the presence of illumination, the photodetector attained a peak QE_{ext} of 69.3%, Riext of 0.26 A/W,

*D** of 8.12×10^{15} cmHz$^{1/2}$/W, and *NEP* of 7.79×10^{-20} W at a bias of −1.0 V. The proposed broadband photodetector exhibited remarkable performance enhancement, which could be attributed to the presence of a strong electric field of 37.83 kV/cm. This substantial field served to accelerate the photogenerated electrons and holes toward the electrodes, thereby enabling efficient collection of the photo-induced carriers. This study highlighted the photodetection of the photodetector in the broadband spectrum region.

While traditional mathematical approaches can be used for design and optimization, they often prove to be time-consuming and computationally demanding, particularly in the context of complex photodetector structures or large datasets. When dealing with large datasets of optoelectronic properties and performance parameters, traditional mathematical analyses are ineffective. However, ML approaches demonstrated superior competence in assessing and forecasting photodetector characteristics, enabling optimized design and enhanced performance. In this work, the DT and ET regression models exhibited outstanding performance in predicting the photodetector performance for test set sizes varying from 40% to 70%, with low values for the MSE, RMSE, and MAE metrics, and an R^2 of 1. This research significantly contributes to the optimization of photodetector designs for broadband applications, encompassing environmental sensing, military surveillance, imaging, and optical communications, through the effective application of ML regression models.

REFERENCES

[1] A. Teke, S. Dogan, F. Yun, M.A. Reshchikov, H. Le, X.Q. Liu, H. Morkoç, S.K. Zhang, W.B. Wang, and R.R. Alfano, "Gallium Nitride/Aluminum Gallium Nitride back-illuminated multiple-quantum-well Schottky barrier ultraviolet photodetectors", *Solid-State Electron.,* vol. 47, no. 8, pp. 1401-1408, 2003.
[http://dx.doi.org/10.1016/S0038-1101(03)00068-6]

[2] L. Gao, C. Chen, K. Zeng, C. Ge, D. Yang, H. Song, and J. Tang, "Broadband, sensitive and spectrally distinctive tin disulfide nanosheet/lead sulfide colloidal quantum dot hybrid photodetector", *Light Sci. Appl.,* vol. 5, no. 7, pp. 1-8, 2016.
[http://dx.doi.org/10.1038/lsa.2016.126] [PMID: 30167178]

[3] X.W. Tong, W.Y. Kong, Y.Y. Wang, J.M. Zhu, L.B. Luo, and Z.H. Wang, "High-performance red-light photodetector based on lead-free bismuth halide perovskite film", *ACS Appl. Mater. Interfaces,* vol. 9, no. 22, pp. 18977-18985, 2017.
[http://dx.doi.org/10.1021/acsami.7b04616] [PMID: 28508625]

[4] D-S. Um, Y. Lee, S. Lim, J. Park, W-C. Yen, Y-L. Chueh, H-J. Kim, and H. Ko, "Indium Gallium Arsenide nanomembrane/silicon van der Waals heterojunction photodiodes with broadband and high photoresponsivity", *ACS Appl. Mater. Interfaces,* vol. 8, no. 39, pp. 26105-26111, 2016.
[http://dx.doi.org/10.1021/acsami.6b06580] [PMID: 27626467]

[5] L. Zeng, S. Lin, Z. Lou, H. Yuan, H. Long, Y. Li, W. Lu, S.P. Lau, D. Wu, and Y.H. Tsang, "Ultrafast and sensitive photodetector based on a platinum diselenide/silicon nanowire array heterojunction with a multiband spectral response from 200 to 1550 nanometers", *NPG Asia Mater.,* vol. 10, no. 4, pp. 352-362, 2018.
[http://dx.doi.org/10.1038/s41427-018-0035-4]

[6] T.H. Flemban, M.A. Haque, I. Ajia, N. Alwadai, S. Mitra, T. Wu, and I.S. Roqan, "A photodetector based on p-type silicon/n-type zinc oxide nanotube heterojunctions with high ultraviolet responsivity", *ACS Appl. Mater. Interfaces,* vol. 9, no. 42, pp. 37120-37127, 2017. [http://dx.doi.org/10.1021/acsami.7b09645] [PMID: 28925680]

[7] X. An, F. Liu, Y.J. Jung, and S. Kar, "Tunable graphene-silicon heterojunctions for ultrasensitive photodetection", *Nano Lett.,* vol. 13, no. 3, pp. 909-916, 2013. [http://dx.doi.org/10.1021/nl303682j] [PMID: 23350824]

[8] R.K. Ratnesh, M.K. Singh, V. Kumar, S. Singh, R. Chandra, M. Singh, and J. Singh, "Mango leaves (*Mangifera Indica*) derived highly fluorescent green graphene quantum dot nanoprobes for enhanced on-off dual detection of cholesterol and iron ions based on molecular logic operation", *ACS Appl. Bio Mater.,* vol. 7, no. 7, pp. 4417-4426, 2024. [http://dx.doi.org/10.1021/acsabm.4c00292] [PMID: 38875229]

[9] G. Li, N. André, B. Huet, T. Delhaye, N. Reckinger, L.A. Francis, L. Liao, J.P. Raskin, Y. Zeng, and D. Flandre, "Enhanced ultraviolet photoresponse in a graphene-gated ultra-thin Si-based photodiode", *J. Phys. D Appl. Phys.,* vol. 52, no. 24, pp. 245101-245107, 2019. [http://dx.doi.org/10.1088/1361-6463/ab12b8]

[10] S. Bansal, K. Prakash, K. Sharma, N. Sardana, S. Kumar, N. Gupta, and A.K. Singh, "A highly efficient bilayer graphene/ZnO/silicon nanowire based heterojunction photodetector with broadband spectral response", *Nanotechnology,* vol. 31, no. 40, pp. 405205-405210, 2020. [http://dx.doi.org/10.1088/1361-6528/ab9da8] [PMID: 32554900]

[11] R.K. Ratnesh, M.K. Singh, and J. Singh, "Enhancing ZnO/Si heterojunction photodetector performance for ultra high responsivity across wide spectral range", *J. Mater. Sci. Mater. Electron.,* vol. 35, no. 11, p. 756, 2024. [http://dx.doi.org/10.1007/s10854-024-12516-5]

[12] S. Bansal, S. Kumar, N. Gupta, and A.K. Singh, "A highly efficient and low noise n-type zinc oxide/p-type silicon heterojunction based ultraviolet detector", in 2021 IEEE Sponsored 2nd Global Conference on Advancement in Technology (GCAT-2021), Nagarjuna College of Engineering & Technology, Bengaluru, Karnataka, India, pp. 1–4, 2021.

[13] S. Bansal, "Simulation and optimization of platinum/zinc oxide Schottky ultraviolet photodetector with aluminum ohmic contact", in 2024 IEEE 3rd International Conference on Innovation in Technology (INOCON), Bangalore, India, pp. 1–6, 2024.

[14] B. Witkowski, "Applications of zinc oxide nanorods and nanowires—A review", *Acta Phys. Pol. A,* vol. 134, no. 6, pp. 1226-1246, 2018. [http://dx.doi.org/10.12693/APhysPolA.134.1226]

[15] A. Garg, and R.K. Ratnesh, "Solar cell trends, and the future: A review", *J. Pharm. Negat. Results,* vol. 13, pp. 2051-2060, 2022. [http://dx.doi.org/10.47750/pnr.2022.13.S06.268]

[16] C.C. Cheng, J.Y. Zhan, Y.M. Liao, T.Y. Lin, Y.P. Hsieh, and Y.F. Chen, "Self-powered and broadband photodetectors based on graphene/ZnO/silicon triple junctions", *Appl. Phys. Lett.,* vol. 109, no. 5, pp. 053501-1, 053501-053505, 2016. [http://dx.doi.org/10.1063/1.4960357]

[17] S. Bansal, A. Das, P. Jain, K. Prakash, K. Sharma, N. Kumar, N. Sardana, N. Gupta, S. Kumar, and A.K. Singh, "Enhanced optoelectronic properties of bilayer graphene/mercury cadmium telluride-based single- and dual-junction photodetectors in the long infrared regime", *IEEE Trans. Nanotechnol.,* vol. 18, pp. 781-789, 2019. [http://dx.doi.org/10.1109/TNANO.2019.2931814]

[18] S. Bansal, K. Sharma, P. Jain, N. Sardana, S. Kumar, N. Gupta, and A.K. Singh, "Bilayer graphene/mercury cadmium telluride-based very long infrared photodetector with superior external quantum efficiency, responsivity, and detectivity", *RSC Advances,* vol. 8, pp. 39579-39592, 2018.

[http://dx.doi.org/10.1039/C8RA07683A] [PMID: 35558011]

[19] S. Bansal, P. Jain, N. Gupta, A.K. Singh, N. Kumar, S. Kumar, and N. Sardana, "A highly efficient bilayer graphene-mercury cadmium telluride heterojunction-based p+-n photodetector for long wavelength infrared", in 2018 IEEE 13th Nanotechnology Materials and Devices Conference (NMDC), 2018, pp. 1–4.

[20] S. Bansal, *Long-wave bilayer graphene/mercury cadmium telluride-based GBp type-II superlattice unipolar barrier infrared detector.* Results in Optics, vol. 12, pp. 100425-1–100425-9, 2023.

[21] S. Bansal, M. Muthukumar, and S. Kumar, *Graphene/mercury cadmium telluride heterojunction-based infrared detectors.* in Handbook of II-VI Semiconductor-Based Sensors and Radiation Detectors: Volume 2, Photodetectors, Springer International Publishing, 2023, pp. 183–202.

[22] S. Bansal, K. Prakash, P. Jain, S. Kumar, N. Gupta, and A.K. Singh, "Photodetectors for security application", In: *Nanoelectronic Devices for Hardware and Software Security.,* A.K. Singh, B. Raj, Eds., 1st ed. CRC Press, 2021, pp. 279-300.

[23] S. Bansal, K. Prakash, N. Sardana, S. Kumar, K. Sharma, P. Jain, N. Gupta, and A.K. Singh, "Bilayer graphene/mercury cadmium telluride-based self-powered mid-wave infrared nBn photodetector", in 2019 IEEE 14th Nanotechnology Materials and Devices Conference (NMDC), Stockholm, Sweden, 2019, pp. 1–4.

[24] S. Bansal, A. Das, K. Prakash, K. Sharma, N. Sardana, G.M. Khanal, S. Kumar, N. Gupta, and A.K. Singh, *Bilayer graphene/mercury cadmium telluride heterojunction-based novel GBn infrared detectors.* vol. 169. Micro and Nanostructures, 2022, pp. 207345-12.

[25] A. Garg, R.K. Ratnesh, R.K. Chauhan, N. Mittal, and H. Shankar, "Current advancement and progress in BioFET: A review", *2022 International Conference on Signal and Information Processing (IConSIP),* 2022, pp. 1-7.
[http://dx.doi.org/10.1109/ICoNSIP49665.2022.10007517]

[26] F. Xia, T. Mueller, Y. Lin, A. Valdes-Garcia, and P. Avouris, "Ultrafast graphene photodetector", *Nat. Nanotechnol.,* vol. 4, no. 12, pp. 839-843, 2009.
[http://dx.doi.org/10.1038/nnano.2009.292] [PMID: 19893532]

[27] T. Mueller, F. Xia, and P. Avouris, "Graphene photodetectors for high-speed optical communications", *Nat. Photonics,* vol. 4, no. 5, pp. 297-301, 2010.
[http://dx.doi.org/10.1038/nphoton.2010.40]

[28] C. Xie, P. Lv, B. Nie, J. Jie, X. Zhang, Z. Wang, P. Jiang, Z. Hu, L. Luo, Z. Zhu, L. Wang, and C. Wu, "Monolayer graphene film/silicon nanowire array Schottky junction solar cells", *Appl. Phys. Lett.,* vol. 99, no. 13, pp. 133113-3, 2011.
[http://dx.doi.org/10.1063/1.3643473]

[29] J. Wei, X. Chu, X.Y. Sun, K. Xu, H.X. Deng, J. Chen, Z. Wei, and M. Lei, "Machine learning in materials science", *InfoMat,* vol. 1, no. 3, pp. 338-358, 2019.
[http://dx.doi.org/10.1002/inf2.12028]

[30] S. Mridha, and D. Basak, "Ultraviolet and visible photoresponse properties of n-type zinc oxide/p-type silicon heterojunction", *J. Appl. Phys.,* vol. 101, no. 8, pp. 083102-083105, 2007.
[http://dx.doi.org/10.1063/1.2724808]

[31] T.F. Zhang, G.A. Wu, J.Z. Wang, Y.Q. Yu, D.Y. Zhang, D.D. Wang, J.B. Jiang, J.M. Wang, and L.B. Luo, "A sensitive ultraviolet light photodiode based on graphene-on-zinc oxide Schottky junction", *Nanophotonics,* vol. 6, no. 5, pp. 1073-1081, 2017.
[http://dx.doi.org/10.1515/nanoph-2016-0143]

[32] Q. Xu, Q. Cheng, J. Zhong, W. Cai, Z. Zhang, Z. Wu, and F. Zhang, "A metal–semiconductor–metal detector based on ZnO nanowires grown on a graphene layer", *Nanotechnology,* vol. 25, no. 5, p. 055501, 2014.
[http://dx.doi.org/10.1088/0957-4484/25/5/055501]

[33] G. Fan, H. Zhu, K. Wang, J. Wei, X. Li, Q. Shu, N. Guo, and D. Wu, "Graphene/silicon nanowire Schottky junction for enhanced light harvesting", *ACS Appl. Mater. Interfaces,* vol. 3, no. 3, pp. 721-725, 2011.
[http://dx.doi.org/10.1021/am1010354] [PMID: 21323376]

[34] L. Duan, F. He, Y. Tian, B. Sun, J. Fan, X. Yu, L. Ni, Y. Zhang, Y. Chen, and W. Zhang, "Fabrication of self-powered fast-response ultraviolet photodetectors based on graphene/zinc oxide:aluminum nanorod-array-film structure with stable Schottky barrier", *ACS Appl. Mater. Interfaces,* vol. 9, no. 9, pp. 8161-8168, 2017.
[http://dx.doi.org/10.1021/acsami.6b14305] [PMID: 28240856]

[35] Z. Bai, J. Liu, F. Liu, and Y. Zhang, "Enhanced photoresponse performance of self-powered ultraviolet–visible photodetectors based on zinc oxide/copper(I) oxide/electrolyte heterojunctions via graphene incorporation", *J. Alloys Compd.,* vol. 726, pp. 803-809, 2017.
[http://dx.doi.org/10.1016/j.jallcom.2017.08.035]

[36] S. Bansal, and P. Jain, "Automatic fault detection in a class of Wheatstone bridge-based transducer using artificial neural networks in Verilog hardware description language", in 2022 2nd International Conference on Advance Computing and Innovative Technologies in Engineering (ICACITE), India, pp. 466–470, 2022.

[37] S. Bansal, "Artificial neural networks supervised learning-based automatic fault detection in a class of Wheatstone bridge-oriented transducers", *2022 IEEE Sponsored Global Conference for Advancement in Technology (GCAT-2022),* 2022pp. 1-7 Bengaluru, Karnataka, India

[38] M. Czajkowski, and M. Kretowski, "The role of decision tree representation in regression problems – An evolutionary perspective", *Appl. Soft Comput.,* vol. 48, pp. 458-475, 2016.
[http://dx.doi.org/10.1016/j.asoc.2016.07.007]

[39] S. Uddin, I. Haque, H. Lu, M.A. Moni, and E. Gide, "Comparative performance analysis of K-nearest neighbour (KNN) algorithm and its different variants for disease prediction", *Sci. Rep.,* vol. 12, no. 1, p. 6256, 2022.
[http://dx.doi.org/10.1038/s41598-022-10358-x] [PMID: 35428863]

[40] M.M. Hameed, M.K. AlOmar, F. Khaleel, and N. Al-Ansari, "An extra tree regression model for discharge coefficient prediction: Novel, practical applications in the hydraulic sector and future research directions", *Math. Probl. Eng.,* vol. 2021, pp. 1-19, 2021.
[http://dx.doi.org/10.1155/2021/7001710]

[41] A.K. Singh, R. Kumar, and S. Jit, Dielectric modulated TFET on SELBOX substrate as a label-free biosensor applications: analytical modeling study and sensitivity analysis., *Physica Scripta,* vol. 99, no. 10, p. 105413, 2024.
[http://dx.doi.org/10.1088/1402-4896/ad7b8b]

[42] S. Bansal, A.K. Rajpoot, G. Chamundeswari, K. Prakash, P.R. Kumar, A.N.Z. Rashed, M.S. Soliman, and M.T. Islam, "Platinum/zinc oxide and platinum/few-layer graphene/zinc oxide Schottky devices with aluminum ohmic contacts using atlas simulation and machine learning", *J. Sci. Adv. Mater. Devices,* vol. 9, pp. 100798-14, 2024.
[http://dx.doi.org/10.1016/j.jsamd.2024.100798]

[43] S. Bansal, A. Jain, S. Kumar, A. Kumar, P.R. Kumar, K. Prakash, M.S. Soliman, M.S. Islam, and M. Tariqul Islam, "Optoelectronic performance prediction of mercury cadmium telluride homojunction photodetector in long wave infrared spectral region using traditional simulations and machine learning models", *Sci. Rep.,* vol. 14, p. 28230, 2024.
[http://dx.doi.org/10.1038/s41598-024-79727-y] [PMID: 39548271]

CHAPTER 3

The Advancements in Imaging Applications for Nanomaterials

Payal Patial[1], **Kanwarpreet Kaur**[2], **Shonak Bansal**[1], **Meet Kumari**[3] and **Mohammed H. Alsharif**[4,*]

[1] *Department of Electronics and Communication Engineering, Chandigarh University, Mohali, Punjab, India*

[2] *Department of Electronics and Communication Engineering, Thapar Institute of Engineering and Technology, Patiala, Punjab, India*

[3] *ECE Department, National Institute of Technology, Delhi, India*

[4] *Department of Electrical Engineering, College of Electronics and Information Engineering, Sejong University, Gwangjin-gu, Seoul-05006, Korea*

Abstract: Nanotechnology has significantly revolutionized numerous fields, including materials science, medicine, and diagnostics, with its ability to manipulate materials at the nanoscale. Central to these advancements is the development of cutting-edge imaging techniques that offer enhanced visualization and characterization of nanomaterials. This paper focuses on recent innovations in imaging technologies for nanomaterial analysis, examining both the evolution of conventional methods and the integration of novel approaches to meet the growing demands of high-resolution and high-sensitivity imaging. Additionally, by providing better resolution, targeting specificity, and signal enhancement, the use of nanoparticles as contrast chemicals in healthcare imaging modalities, such as magnetic resonance imaging, computed tomography, ultrasonography, PET, and SPECT, has improved illness detection and diagnosis.

Nanoparticles' ability to exploit passive and active targeting mechanisms, such as the enhanced permeability and retention (EPR) effect, and functionalization with ligands for receptor-specific targeting, has revolutionized imaging in oncology, cardiovascular diseases, and neurological disorders. This chapter further highlights the development of multimodal imaging platforms, where nanoparticles are designed to integrate multiple imaging modalities, such as MRI-fluorescence or PET-CT, providing a more holistic diagnostic capability. These multimodal platforms enable the simultaneous acquisition of anatomical and molecular data, improving diagnostic accuracy and allowing for real-time monitoring of therapeutic responses. The integration of machine learning

* **Corresponding author Mohammed H. Alsharif:** Department of Electrical Engineering, College of Electronics and Information Engineering, Sejong University, Gwangjin-gu, Seoul-05006, Korea;
E-mail: malsharif@sejong.ac.kr

Ramesh Kumar, Anupma Gupta, Manish Singla, Arjun Puri, Ashish Kumar Singh & Vipan Kumar (Eds.)

algorithms for image analysis and interpretation is another area of exploration, with the potential to automate and enhance the accuracy of nanomaterial imaging. While nanoparticle-based imaging holds great promise, challenges remain in ensuring their biocompatibility, biodistribution, and long-term safety. This chapter emphasizes the need for continued research into the optimization of nanoparticle design, surface functionalization, and toxicity profiles to advance their clinical translation. By reviewing the current state of imaging technologies and their applications, this research outlines the critical role that advanced imaging techniques play in pushing the frontiers of nanotechnology, ultimately paving the way for significant breakthroughs in science, medicine, and engineering.

Keywords: CT scan, EPR, MRI, PET scan, Ultrasound.

INTRODUCTION

An essential component of the healthcare system is the prompt identification and diagnosis of illnesses, including cancer. Studies have shown significant improvements in survival rates for patients who benefit from early detection. For instance, gastrointestinal cancer patients with early detection have demonstrated a two-year survival rate of 92.3%, compared to 33.3% for those without [1]. Similarly, breast cancer patients with early detection have seen a reduction in ten-year mortality rates by 17-28% [2]. Advanced diagnostic techniques in healthcare rely heavily on medical imaging to identify diseases in their early stages and monitor treatment efficacy. The field encompasses a range of sophisticated technologies, each offering unique insights into the human body. These include radiographic methods like traditional X-rays and the more detailed computed tomography (CT) scans [3]. Magnetic resonance imaging (MRI) provides high-resolution images of soft tissues, while ultrasound (US) uses sound waves for real-time visualization [4]. Nuclear medicine techniques such as positron emission tomography (PET) and single photon emission computed tomography (SPECT) offer functional imaging capabilities. Additionally, fluorescence imaging has emerged as a valuable tool in certain medical applications [5]. Together, these diverse imaging modalities form the cornerstone of modern diagnostic medicine, enabling healthcare professionals to visualize and analyze internal structures and physiological processes with unprecedented clarity.

In order to provide more accurate structural and physiological details, contrasting chemicals are used to distinguish between normal cells and pathological lesions [6]. Despite advancements in medical imaging, current contrast agents face limitations due to hardware constraints. For instance, contrast-enhanced CT can only detect liver tumors down to 3 mm in size [7]. Most conventional contrast agents are small molecules, which present challenges such as quick metabolic clearance, non-targeted distribution throughout the body, and potential safety issues. The efficacy and clinical usefulness of conventional imaging contrast materials are limited by these considerations taken together. These drawbacks show how better contrast agents are required to improve early illness identification and detection. Nanomaterials have emerged as promising candidates for improving medical imaging and detection capabilities. Their unique properties enable passive, active, and physical targeting mechanisms [8]. The increased permeability and retention (EPR) action of nanoparticles can cause them to concentrate in tumour tissues, raising the local amounts of contrast chemicals. The size of nanoparticles is a critical factor in their effectiveness for tumor imaging [9].

It influences several key aspects of their behavior in the body, including biodistribution patterns, duration of circulation in the bloodstream, uptake by cells, ability to penetrate tumor tissues, and overall targeting efficiency [10]. These size-dependent characteristics make nanoparticles particularly interesting for developing advanced imaging contrast agents. Research has shown that nanoparticles smaller than 10 nm, which is the average size of renal filtration pores, are quickly eliminated through the renal excretion system [11]. Conversely, nanoparticles larger than 100 nm are more easily recognized by macrophages and tend to accumulate in organs associated with the mononuclear phagocyte system (MPS), such as lymph nodes, liver, spleen, and lungs [12]. Multiple reviews have consistently reported that nanoparticles with sizes ranging from 10 to 60 nm demonstrate enhanced cellular uptake [13]. This size range appears to offer a balance between avoiding rapid renal clearance and minimizing recognition by the MPS, making it particularly promising for tumor imaging applications [14]. These findings highlight the importance of carefully considering nanoparticle size when designing contrast agents for medical imaging, particularly in the context of tumor detection and diagnosis. Nanoparticles in imaging have evolved beyond passive targeting, incorporating active targeting strategies through surface modifications [15]. By attaching specific ligands, these nanoparticles can bind to particular receptors, enhancing their accumulation at target sites within lesions. Two approaches, such as CT imaging of prostate cancer, and gold nanoparticles, have been functionalized with RNA aptamers that recognize prostate-specific membrane antigens. This modification resulted in improved contrast density at tumor sites [16].

For MRI applications in lung cancer, superparamagnetic iron oxide (SPIO) nanoparticles have been coupled with anti-EGFR antibodies. This design allows for targeted visualization of lung tumors expressing high levels of EGFR [17]. These developments show how customised nanoparticle architectures can greatly improve therapeutic imaging's specificity and effectiveness across a range of modalities and illness types. Nanoparticle-based imaging agents can be enhanced through various targeting approaches to improve their precision and efficacy. Antibodies and their fragments, known for their high specificity and strong affinity to antigens, help to increase the concentration of nanoparticles at target sites. Aptamers, which are short DNA or RNA sequences with high receptor affinity, offer advantages, such as smaller size, easier synthesis, and lower immunogenicity; however, they are limited by rapid clearance and susceptibility to degradation by nucleases. Peptides provide chemical stability and reduced immunogenicity, with the RGD peptide being a notable example for targeting integrins. Additionally, other molecules like transferrin, folic acid, and biotin can be used for targeted delivery [18]. Each of these strategies comes with its own set of advantages and limitations, with the optimal method depending on the specific imaging application, target tissue, and the properties required for the nanoparticles. Beyond active and passive targeting, researchers are also investigating physical targeting techniques, such as the use of external stimuli, to further refine nanoparticle localization and improve imaging performance. These approaches involve guiding nanoparticles to target sites and controlling release processes through external sources or fields, as seen in photothermal and magnetic hyperthermia therapies. For instance, dual-stimuli responsive nanoprobes sensitive to acidic pH and reduction have been developed for enhanced CT imaging of tumors [19].

Nanoparticle-based contrast agents represent a significant advancement in biomedical imaging, offering extended blood circulation times crucial for enhanced imaging. Various surface modifications, including PEG encapsulation and zwitterionic alterations, have been developed to prolong circulation and improve stability. These nanoscale agents excel in early tumor detection, serving as versatile contrast enhancers and biomarker detectors. Nanoparticle-based imaging agents include high surface area for functionalization, improved toxicity profiles, functional visualization of biological processes, extended plasma circulation, and customizable physical properties [20]. Their potential in biomedical applications is reflected in the rapid growth of related publications, signaling a transformative impact on diagnostic imaging across various modalities. Nanoparticles have emerged as versatile tools across multiple biomedical imaging techniques. This review explores their utilization in fluorescence imaging, magnetic resonance imaging (MRI), computed tomography (CT), ultrasound (US), positron emission tomography (PET), and single-photon emission computed tomography (SPECT) [21, 22]. By examining the structural characteristics of various nanoparticles, we highlight their specific applications in disease detection and monitoring. The unique properties of these nanoscale materials enable enhanced contrast, improved sensitivity, and targeted imaging capabilities, potentially revolutionizing diagnostic approaches across a spectrum of medical conditions. Our primary focus is on examining the structural characteristics of nanoparticles and their relevant applications in biomedical imaging. This includes their use in the imaging of diseases. For each imaging modality, we will explore frequently researched examples to illustrate the potential and current state of nanoparticle-based imaging techniques.

NANOPARTICLE-ENHANCED COMPUTED TOMOGRAPHY: A NEW ERA IN DIAGNOSTIC IMAGING

CT Imaging: Strengths and Constraints

The sophisticated diagnostic imaging technique known as computed tomography (CT) creates cross-sectional and 3-D anatomical pictures by taking advantage of tissue thickness and differential X-ray attenuation. CT has rapidly displaced conventional plain-film radiography in clinical practice due to its superior examination speed, cost-effectiveness, enhanced efficiency, and improved spatial resolution. Nonetheless, this advancement is accompanied by increased ionizing radiation exposure to patients. CT contrast agents are essential for differentiating tissues with similar attenuation coefficients, thereby improving diagnostic accuracy. Currently, intravenous CT contrast agents predominantly use iodine-based formulations. Despite their widespread application, these agents are associated with limitations such as rapid systemic clearance, potential nephrotoxicity, with negative consequences, such as anaphylaxis [23]. To go over these restrictions and broaden the scope of CT imaging, researchers have explored nanosized contrast agents. Nanoparticulate systems offer several advantages over traditional iodinated agents, including prolonged circulation times that allow for extended imaging windows, reduced renal toxicity through alternative clearance mechanisms, potential for targeted delivery to specific tissues or pathologies, and enhanced contrast-to-noise ratios, which enable lower radiation doses. The development of nanoparticle-based CT contrast asignifies a substantial advancement in molecular imaging, with the potential to provide more precise diagnoses, improve treatment planning, and enhance monitoring of therapeutic responses [24].

Structural and Compositional Classification of Nanoparticle-based CT Contrast Agents

Computed tomography (CT) contrast agents, according to the makeup of nanoparticles, can be categorized into two primary types, each offering distinct advantages for diagnostic imaging applications. Iodine-based contrast agents are included in the first group, which leverage the well-established clinical history and safety profile of iodine. In these formulations, nanoparticles act as delivery vectors for iodine, typically using a core-shell architecture [24]. The iodine is encapsulated within the nanoparticle core, as demonstrated by liposomal iodine formulations. This approach enhances iodine concentration per particle, potentially improving contrast efficacy while addressing limitations associated with free iodine administration. Using nanoparticles made of high-atomic-number metals with excellent X-ray absorption capabilities, the second group includes metal-based contrast agents. Au, Ta_2O_5, and ZrO_2 nanoparticles are a few examples [25]. Those metal-based nanostructures may employ fundamental, core-shell, or vector layouts to mimic the physical makeup of magnetic resonance imaging (MRI) contrast agents [25]. These metal-based nanostructures can be structured similarly to magnetic resonance imaging (MRI) contrast agents, employing basic, core-shell, or vector designs. In the basic structural configuration, gold nanoparticles are the most common, showing considerable potential for use as CT imaging contrast agents. Core-shell designs offer flexibility in metal incorporation [26]. Nanoparticles with gold nanosphere cores surrounded by mesoporous silica shells filled with indocyanine, along with those that contain zirconium dioxide in the outermost layer and doxorubicin and polypyrrole (PPy) inside, are noteworthy examples. Vector designs utilize metal elements as surface materials, exemplified by poly(acrylic acid)-bridged gadolinium nanoparticles functionalized with surface gold elements [27]. This approach allows for the integration of multiple functionalities, combining both diagnostic and therapeutic capabilities within a single nanoplatform. These diverse nanoparticle-based CT contrast agent designs offer enhanced imaging capabilities, extended circulation times, and targeted delivery potential, promising significant advancements in diagnostic accuracy and treatment monitoring [28, 29]. Fig. (**1**) shows the different Structures for CT contrast agents.

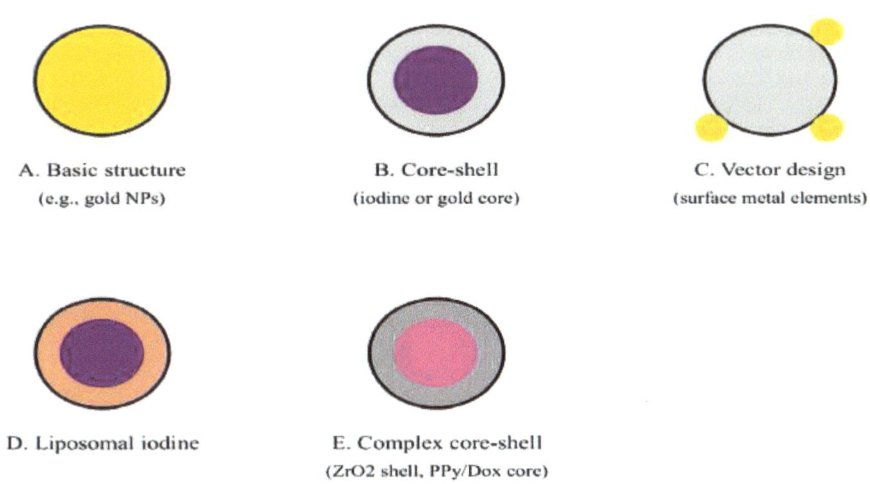

Fig. (1). Reflects the different Structures for CT contrast agents.

Nanoparticles: Innovations in Biomedical Imaging

Nanosized CT contrast agents have emerged as versatile tools in biomedical imaging, leveraging their unique properties to enhance the visualization of various biological processes. These nanoparticles, characterized by their size, composition, and functionalization, offer significant advantages in CT imaging by providing robust contrast, targeted delivery, and the potential for multimodal imaging applications. Blood Flow Imaging benefits from gold nanoparticles encapsulated within red blood cells, which offer high contrast for visualizing blood flow dynamics due to their effective CT attenuation properties. Tumor Vessel Evaluation is achieved using liposomal iodine, which enhances CT imaging through long circulation times and strong contrast, allowing detailed assessment of tumor vasculature [30]. For Prostate Cancer Imaging, gold nanoparticles functionalized with prostate-specific membrane antigen (PSMA) aptamers enable targeted detection of prostate cancer cells, improving tumor characterization and potential early detection. Tumor Imaging and Drug Distribution Monitoring utilizes zirconium dioxide nanoparticles to visualize tumor tissues and track drug distribution, providing insights into both tumor morphology and therapeutic efficacy. The key advantages of nanosized CT contrast agents include their ability to concentrate high atomic number elements in a small volume, thereby enhancing contrast. Surface modifications with specific ligands or antibodies enable targeted imaging, allowing for more precise detection of disease-specific markers. Additionally, these nanoparticles offer the potential for multimodal imaging, where CT can be combined with other imaging techniques to provide comprehensive diagnostic information [31]. The integration of diagnostic and therapeutic functions, known as theranostics, is another promising application, potentially allowing simultaneous disease detection and treatment. The accompanying Table **1** summarizes a range of CT nanoimaging agents from the literature, highlighting their diverse compositions, sizes, and applications. Notably, many of these agents have progressed to *in vivo* studies, underscoring their potential for clinical application. As research advances, further refinements in nanoparticle design are expected to enhance the sensitivity, specificity, and safety profiles of CT contrast agents. This progress holds promise for improving diagnostic capabilities and patient outcomes through earlier and more accurate detection of various medical conditions [32].

Table 1. A range of CT nanoimaging agents from the literature, highlighting their diverse compositions, sizes, and applications.

S.No.	Nanoparticle Composition	Size (nm)	Biomedical Application
1	Gold nanoparticles (in RBCs)	15-20	Blood flow imaging
2	Liposomal iodine	100-150	Tumor vessel evaluation

(Table 1) cont.....

S.No.	Nanoparticle Composition	Size (nm)	Biomedical Application
3	Gold nanoparticles with PSMA-aptamer	10-30	Prostate cancer imaging
4	Zirconium dioxide nanoparticles	50-100	Tumor imaging and drug distribution monitoring
5	Tantalum oxide nanoparticles	30-50	Lymph node mapping
6	Gold nanorods	40-60	Tumor margin delineation
7	Bismuth sulfide nanoparticles	20-40	Gastrointestinal tract imaging
8	Iodine-loaded polymeric nanoparticles	80-120	Liver and spleen imaging
9	Gadolinium-based nanoparticles	02-05	Vascular imaging
10	Hafnium oxide nanoparticles	50-70	Bone imaging

CT Imaging: Notable Nanoparticle Applications

Gold nanoparticles (AuNPs) have gained recognition as versatile agents for biomedical imaging, particularly in computed tomography (CT), due to their superior X-ray absorption properties. AuNPs exhibit significantly higher absorption efficiency compared to iodine-based agents, bone, and soft tissues (5.16 cm^2/g *vs.* 1.94, 0.186, and 0.169 cm^2/g at 100 keV, respectively), making them highly effective CT contrast agents. Their biocompatibility and biosafety profile, coupled with customizable sizes and shapes optimally stable within the 4 to 30 nm diameter range, enhance their cell specificity and functionality. Unlike conventional iodine-based contrast agents, AuNPs offer improved contrast resolution while minimizing the risks of renal toxicity and osmotic damage. AuNPs have been successfully used as cellular tracers for monitoring tumor progression through X-ray imaging and for tracking blood flow *via* labeled erythrocytes. Their utility in active and passive targeting mechanisms further enhances cancer detection and improves radiotherapy outcomes, making them valuable tools in both diagnosis and treatment. In parallel, iodine-based liposomal nanotechnology has been developed to address limitations of traditional iodinated agents. Liposomal iodine nanoparticles, typically sized between 100 - 400 nm, offer reduced osmotic pressure, extended circulation times, and enhanced contrast properties. These nanoparticles are capable of targeting lesions through cellular uptake, surface-specific binding, or by leveraging the enhanced permeability and retention (EPR) effect, thus expanding the scope of nanoparticle-based contrast agents in medical imaging. The incorporation of these nanoparticle-based CT contrast agents into clinical practice promises advancements in diagnostic precision and broadens the application potential of CT imaging across various medical fields [33, 34].

NANOPARTICLE-ENHANCED FLUORESCENCE IMAGING: A NEW ERA IN DIAGNOSTIC IMAGING

Fluorescence Imaging: Strengths and Constraints

Fluorescence Imaging provides excellent spatial resolution for diagnosing diseases at the microscopic level. However, it is hindered by challenges, such as limited tissue penetration, autofluorescence, and light scattering. Nanoparticles offer promising solutions to these limitations. They can be tailored to carry larger amounts of fluorescent dyes, which boosts signal intensity. Additionally, their structure can be optimized to prevent quenching of NIR fluorescence. Nanoparticles can also use both passive and active targeting techniques to concentrate more effectively in diseased tissues, enhancing local fluorescence. Their prolonged circulation time further facilitates higher accumulation in target areas. Moreover, nanoparticles are capable of photon upconversion, which reduces issues like blinking and photobleaching. These unique features have driven extensive research into nanoparticle-based fluorescent platforms, aiming to surpass the limitations of traditional fluorescence imaging and improve its effectiveness in clinical disease diagnosis and detection [35, 36].

Structural and Compositional Innovations

Nanoparticle-based fluorescence imaging technologies offer diverse strategies to enhance imaging precision and efficiency. One approach involves surface-labeled fluorophores, where fluorescent molecules are attached to the nanoparticle's surface. For example, plasmonic or magnetic nanoparticles combined with Cy3-modified S6 aptamers increase local fluorophore concentrations, improving imaging. Another design uses core-shell structures, with fluorescent dyes encapsulated in the nanoparticle core and targeting molecules on the outer shell. These systems, such as micelles, dendrimers, and quantum dots, enhance the solubility of hydrophobic dyes and protect them from degradation. Quenched fluorescent dye nanoparticles contain concentrated, inactive dyes that are released and activated at target sites, boosting resolution and improving the target-t--background ratio, which is useful in detecting protease activity, tumors, and inflammation. A related method involves separate quenching molecules that suppress fluorescence until specific enzymes, like DNAzymes, activate the signal. Examples include Zn-Enz/AuNP-FAM/BHQ-1 nanoparticles. Förster Resonance Energy Transfer (FRET) further enhances fluorescence imaging by transferring energy between donor and acceptor molecules. In particular, upconversion nanoparticles (UCNPs) are valuable for their use of NIR light, which reduces tissue damage, allows deeper penetration, and improves the signal-to-background ratio. These nanoparticles convert NIR excitation into visible fluorescence, offering significant advantages for medical imaging [37 - 39]. Fig. (**2**) emphasizes the different nanoparticle designs for fluorescence imaging.

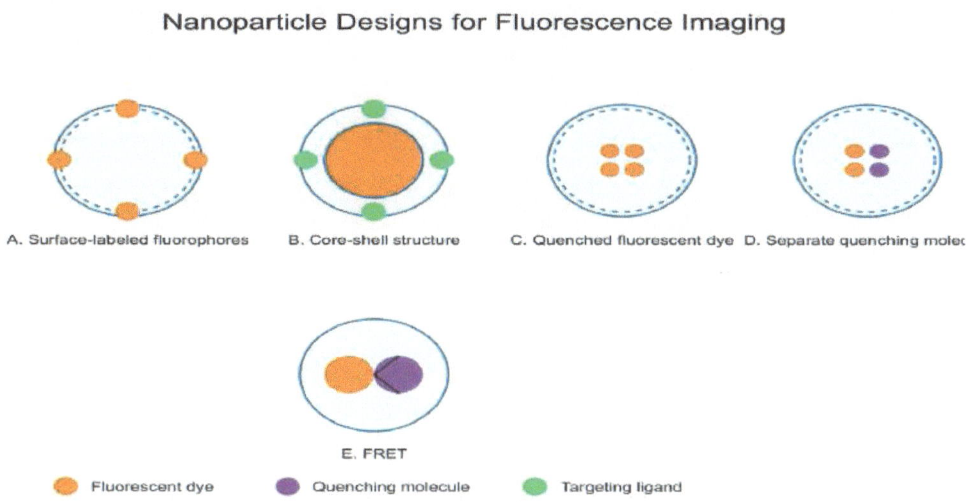

Fig. (2). Different Nanoparticle designs for fluorescence imaging.

Nanoparticles: Innovations in Biomedical Engineering

Nanoparticle-based fluorescence imaging has emerged as a powerful tool in biomedical research, offering versatility across a wide range of applications, from molecular-level studies to organism-wide diagnostics. These applications are categorized into several domains, including gene and protein detection, enzyme activity evaluation, and cellular tracking. In clinical settings, this technology has been employed for early disease diagnosis, tumor research, and real-time monitoring of therapeutic responses. The ability of nanoparticles to enhance imaging precision and sensitivity, combined with their adaptability to different biological systems, highlights their significant potential for advancing both research and clinical diagnostics. The systematic exploration of nanoparticle composition, size, and imaging agents, as summarized in Table **2**, provides insights into the relationship between nanoparticle properties and their specific applications, demonstrating the ongoing evolution of this field from *in vitro* studies to *in vivo* implementations [40, 42].

Table 2. Insights into the relationship between nanoparticle properties and their specific applications.

S.No.	Nanoparticle Composition	Imaging Agent	Size (nm)	Biomedical Application
1	Gold nanoparticles	Fluorescein	15-20	Gene detection
2	Silica nanoparticles	Rhodamine B	50-70	Protein analysis
3	Quantum dots (CdSe/ZnS)	Intrinsic	05-10	Enzyme activity evaluation
4	Lanthanide-doped nanoparticles	Eu^{3+}, Tb^{3+}	30-50	Element tracing
5	Iron oxide nanoparticles	Cy5.5	20-30	Cell tracking
6	Polymer nanoparticles	ICG	80-100	Early-stage disease diagnosis
7	Carbon nanotubes	NIR-II dyes	1-2 (diameter)	Tumor-related research
8	Upconversion nanoparticles	Yb^{3+}, Er^{3+}	40-60	Monitoring real-time therapeutic effects

Fluorescence Imaging: Notable Nanoparticle Applications

The diverse types of nanoparticles are employed in fluorescence imaging, with their structural properties, advantages, and biomedical applications. Quantum dots (QDs), semiconductor nanocrystals with core-shell architectures measuring 1 to 10 nm, display size-dependent band gaps, which allow for broad absorption spectra and tunable fluorescence emission. Operating through Förster Resonance Energy Transfer (FRET), QDs offer improved tissue penetration and minimized background autofluorescence. Their photostability, narrow emission bands, and high molar extinction coefficients render them superior to traditional organic dyes. However, challenges such as cytotoxicity due to the release of heavy metal ions and their inherent hydrophobicity pose limitations for biological applications. Gold nanoparticles (AuNPs), ranging in size from a few to several hundred nanometers, exhibit high fluorescence quenching efficiency, enabling precise "on-off" control of fluorescence signals. AuNPs offer strong photostability, biocompatibility, and ease of surface functionalization, making them suitable for various imaging modalities. Similarly, graphene oxide and carbon nanotubes also demonstrate efficient quenching effects, with additional advantages such as low cost, high sensitivity, and good solubility across a range of solvents. Dark quenchers, non-fluorescent molecules that work synergistically with nanoparticles and fluorophores, play a pivotal role in applications such as gene and protein detection, early tumor detection, and enzyme activity assays. Commercial dark quenchers include Dabsyl, Black Hole Quenchers (BHQ), and Iowa Black, among others. Upconversion nanoparticles (UCNPs), commonly doped with lanthanides, are emerging as promising tools in fluorescence imaging. These nanoparticles, composed of elements like La, Nb, Gd, Y, and Yb as absorbers and Tm, Er, and Ho as emitters, absorb near-infrared light and emit higher-energy visible light.

UCNPs, with sizes ranging from 8 to 550 nm, are highly photostable and biocompatible, offering bright fluorescence signals. They have demonstrated utility in protein detection, cell imaging, and cancer diagnostics. However, similar to QDs, the hydrophobic nature of UCNPs necessitates surface modification to enhance their compatibility with biological environments [43 - 45].

NANOPARTICLE-ENHANCED MRI IMAGING: A NEW ERA IN DIAGNOSTIC IMAGING

MRI Imaging: Strengths and Constraints

Magnetic Resonance Imaging (MRI) is a widely utilized imaging modality in clinical diagnostics, leveraging the principles of proton spin dynamics within an external magnetic field, which are excited by a radio frequency pulse. By detecting the nuclear magnetic resonance signals from protons in biological tissues, MRI offers high spatial and temporal resolution along with exceptional intrinsic soft-tissue contrast. Its capability to generate detailed three-dimensional anatomical tomographic information further enhances its diagnostic utility. Importantly, MRI does not involve ionizing radiation or radiotracers, making it a safer alternative compared to other imaging techniques such as computed tomography (CT). However, MRI does have certain limitations, including its relatively high cost, longer imaging times, and susceptibility to motion artifacts, as well as interference from metallic foreign bodies or implants. To address some of these challenges, MRI contrast agents play a critical role in improving lesion detection and facilitating the differentiation between pathological and healthy tissues. These agents significantly enhance image clarity and diagnostic accuracy in a range of clinical applications [46, 47].

Structural and Compositional Innovations

Magnetic Resonance Imaging (MRI) is a widely utilized diagnostic technique that leverages the nuclear magnetic resonance (NMR) signals of hydrogen protons in biological tissues under the influence of a strong magnetic field. To further enhance the quality of MRI scans, contrast agents are employed, which can be broadly classified into T1, T2, and dual-weighted agents. T1-weighted agents typically involve paramagnetic materials, such as gadolinium-based chelates, which reduce the longitudinal relaxation time (T1) of surrounding protons, resulting in brighter images on T1-weighted sequences. These are often referred to as "positive" contrast agents due to their brightening effect on the image. In contrast, T2-weighted agents consist of superparamagnetic materials, such as superparamagnetic iron oxide nanoparticles (SPIONs), which shorten the transverse relaxation time (T2), leading to signal decay and darker regions in T2-

weighted images, hence functioning as "negative" contrast agents. Emerging technologies also feature dual-weighted agents, such as gadolinium oxide nanoparticles, that simultaneously offer both T1 and T2 contrast, allowing for more versatile imaging. To overcome the limitations inherent in proton-based MRI, heteronuclear MRI approaches have been developed, focusing on nuclei like F, C, P, Na, and O. Among these, F-based contrast agents have gained significant attention due to their applicability in tracking cellular processes, tumor labeling, and pulmonary imaging, thanks to the absence of background signals from endogenous fluorine atoms in the body. The structural design of these MRI contrast nanoparticles is critical in determining their effectiveness. Various designs, such as surface-decorated magnetic nanoparticles, core-shell architectures, vector-based structures, and mixed material systems, have been employed to tailor properties for specific imaging modalities. Surface-functionalized nanoparticles, for instance, can enhance target-specific delivery through ligand-receptor interactions, while core-shell designs offer improved biosafety and stability, encapsulating the magnetic core with a biocompatible shell. These structural innovations enable nanoparticles to serve dual or even multimodal functions, incorporating both diagnostic and therapeutic capabilities, commonly referred to as theranostic agents. In conclusion, nanoparticle-based contrast agents, with their advanced structural flexibility and ability to enhance MRI sensitivity and specificity, represent a significant advancement in diagnostic imaging, with potential implications across oncology, neurology, and cardiovascular medicine [48, 49]. Fig. (**3**) represents the structural designs of MRI contrast nanoparticles.

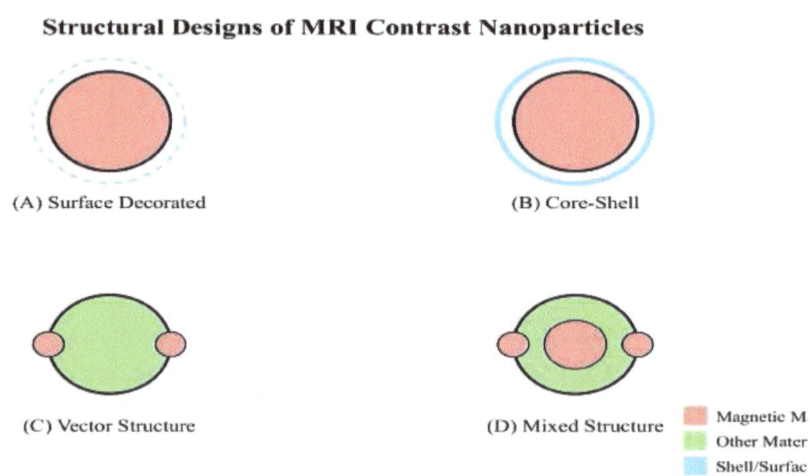

Fig. (3). Represents the structural designs of MRI contrast nanoparticles.

Nanoparticles: Innovations in Biomedical Imaging

Nanoparticle-based contrast agents have significantly advanced magnetic resonance imaging (MRI), enabling high-resolution visualization at the molecular, cellular, and organ levels. The effectiveness of these contrast agents is largely governed by the targeting approach utilized, which can be broadly classified into active and passive strategies. Active targeting is achieved through surface functionalization, which enhances the biodistribution of nanoparticles by attaching specific antibodies, ligands, or peptides, particularly for tumor diagnosis. This method improves specificity in imaging by directing nanoparticles to desired biological targets. In contrast, passive targeting is size-dependent, with nanoparticles typically ranging from several nanometers to hundreds of nanometers. This approach leverages the enhanced permeability and retention (EPR) effect, commonly exploited in tumor imaging, allowing nanoparticles to accumulate in regions with compromised vasculature, such as tumors. Beyond tumor imaging, nanoparticle-based MRI contrast agents are applied in various non-specific cellular uptake mechanisms. These applications include blood pool contrast enhancement, detection of inflammation, identification of malignant lymph nodes, tracking of mesenchymal stem cells, and monitoring of islet transplantation. The literature presents a broad spectrum of nanoparticle-based MRI contrast agents, characterized by diverse compositions, sizes, and biomedical applications. Table **3** provides a comprehensive overview of these agents, detailing their material composition, imaging modalities, size parameters, target applications, and experimental validation, whether conducted *in vivo* or in vitro. The ongoing development of these nanoparticle-based agents demonstrates their immense potential to enhance MRI diagnostic accuracy across a wide range of medical applications, including oncology, inflammatory diseases, and regenerative medicine [50, 51].

Table 3. Overview of agents, their material composition, imaging modalities, size parameters, and target applications.

S.No.	Nanoparticle Composition	Imaging Agent	Size (nm)	Biomedical Application
1	SPIO-coated liposomes	T2-weighted	100-200	Tumor imaging
2	Gd-DTPA-loaded dendrimers	T1-weighted	05-Oct	Blood pool imaging
3	Manganese oxide nanoparticles	T1-weighted	20-40	Stem cell tracking
4	Gadolinium oxide nanoparticles	Dual T1/T2	02-05	Tumor diagnosis
5	19F-perfluorocarbon nanoparticles	19F-MRI	200-300	Inflammation imaging
6	Iron oxide-gold hybrid nanoparticles	T2-weighted	50-100	Photothermal therapy
7	Silica-coated iron oxide nanoparticles	T2-weighted	30-50	Lymph node mapping
8	Gd-doped upconversion nanoparticles	T1-weighted	10-30	Multimodal imaging

MRI imaging: Notable Nanoparticle Applications

Nanoparticle-based MRI contrast agents offer a wide range of advantages in medical imaging, with each type designed to enhance specific imaging capabilities. Gadolinium-based agents, known for producing bright T1-weighted images, have evolved with nanotechnology to improve their performance and reduce toxicity risks by using chelates or surface conjugation. Ferric oxide-based agents, which generate dark T2-weighted images, have been widely used in molecular imaging and cell tracking, though concerns about toxicity have paused some clinical applications. Manganese-based nanoparticles, with their ability to shorten T1 relaxation times, are emerging as alternatives, offering less toxicity and potential applications in cancer diagnosis and lymph node imaging. Additionally, fluorine-based agents like 19F nanoparticles bring unique advantages, such as producing no background signal, allowing for precise tumor detection and tracking of therapeutic responses. The development of these contrast agents has also paved the way for theranostic applications, combining diagnosis and therapy, while ongoing research focuses on enhancing the biocompatibility and reducing the risks of toxicity, thereby advancing their potential for clinical translation [52, 53]. Fig. (**4**) showcases the comparison of MRI contrast agents.

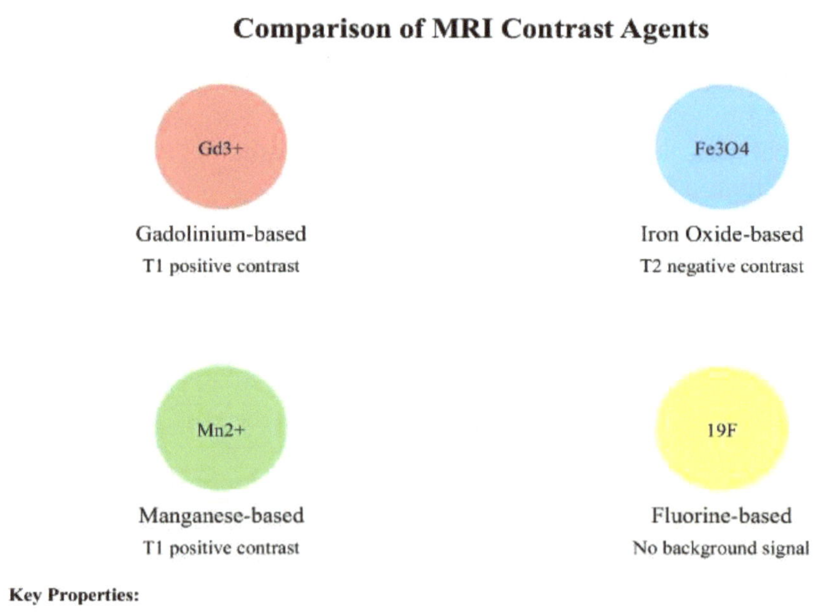

Fig. (4). Comparison of key properties of MRI contrast agents.

NANOPARTICLE-ENHANCED MULTIPODALITY IMAGING: A NEW ERA IN DIAGNOSTIC IMAGING

Multimodality Imaging: Strengths and Constraints

In the realm of medical imaging, individual modalities each present distinct advantages and limitations. Fluorescence imaging, for instance, offers superior spatial resolution but is constrained by limited tissue penetration. Conversely, Positron Emission Tomography (PET) excels in sensitivity and provides functional information with excellent tissue penetration, though it suffers from relatively poor spatial resolution. Computed Tomography (CT) is notable for its high spatial resolution but involves ionizing radiation, which poses a potential risk. To address these limitations and harness the strengths of various imaging techniques, multimodality imaging has emerged as a promising approach. This technique utilizes multifunctional nanoparticles as contrast agents to enhance diagnostic capabilities. For example, the Cu-Fe-RGD-PEG-MNP nanoparticle integrates photoacoustic, PET, and MRI imaging, while fluorescein isothiocyanate (FI) conjugated PEGylated gold nanoparticles combine fluorescence and CT imaging. These multimodal systems offer significant potential benefits, including reduced contrast agent doses and improved targeting capabilities. However, the adoption of multimodality imaging faces several challenges. The lack of standardized criteria due to the diverse designs of these systems, coupled with the need for specialized equipment capable of integrating multiple imaging modules, highlights the considerable hurdles that remain. Despite its promise, multimodality imaging is still in its nascent stages and will require further development before it can be routinely implemented in clinical practice [54, 55].

Structural and Compositional Innovations

Multifunctional nanoplatforms for multimodality imaging represent a significant advancement in diagnostic imaging, yet their synthesis is notably more complex compared to single-modality nanoplatforms. These advanced nanoparticles come in various forms, each designed to support multiple imaging modalities. For instance, single agents such as gadolinium oxide (Gd2O3) can serve both CT and MRI imaging purposes, while other platforms may integrate distinct imaging agents, such as iron oxide combined with NIR830 dye, to facilitate both MRI and fluorescence imaging. The diversity of nanostructures available for these multifunctional platforms allows for a broad range of imaging combinations, including MRI-fluorescence, MRI-PET, CT-fluorescence, and more complex setups like MRI-CT-fluorescence or US-CT-MRI-fluorescence. This versatility underscores the potential of these nanoplatforms to provide a more comprehensive diagnostic picture by leveraging the strengths of different imaging modalities simultaneously. Overall, the development and application of multifunctional nanoplatforms aim to overcome the limitations inherent in individual imaging techniques, thus enhancing diagnostic accuracy and clinical utility [56, 58].

Nanoparticles: Innovations in Biomedical Imaging

The growing interest in multimodality nanoparticle-based imaging highlights its promising potential for enhancing diagnostic accuracy across various medical applications. One of the most significant areas of interest is early tumor detection. The capability of these nanoparticles to integrate multiple imaging modalities suggests they could offer superior sensitivity and specificity, allowing for the identification of tumors at earlier stages than traditional single-modality imaging techniques. Beyond oncology, researchers are exploring the use of multimodality nanoparticles in several other critical body systems. Applications are being investigated for the cardiovascular system, renal system, neural system, and lymphatic system. This broad exploration underscores the versatility of multimodality imaging, as it has the potential to provide more comprehensive diagnostic information across a range of physiological and anatomical contexts. By leveraging the strengths of different imaging techniques within a single nanoparticle-based system, these advanced imaging tools could facilitate earlier disease detection and more precise, personalized treatment strategies. This aligns with the broader trend in medical imaging towards developing more detailed and accurate diagnostic methods, ultimately aiming to improve patient outcomes and enhance overall healthcare [59, 60].

Multimodality Imaging: Notable Nanoparticle Applications

Photoacoustic imaging represents a sophisticated hybrid technique that combines optical and ultrasound imaging to achieve high contrast and spatial resolution while penetrating deeper into biological tissues. This imaging modality operates by converting absorbed optical energy into ultrasound emissions within the tissue. Its integration with photothermal therapy is often pursued due to the heat produced during the imaging process. A key component in photoacoustic imaging is the use of contrast agents, which can be either endogenous compounds like hemoglobin and melanin or, more recently, a diverse array of exogenous nanoparticles. Exogenous nanoparticles are the focus of considerable research and development due to their potential to significantly enhance imaging capabilities. These nanoparticles fall into several categories: inorganic materials such as metal nanoparticles (*e.g.*, gold, silver, copper) and carbon nanomaterials (*e.g.*, carbon nanotubes, graphene); organic materials including small molecule dyes (*e.g.*, silicon naphthalocyanine, indocyanine green); and microbubble-producing nanoparticles (*e.g.*, CO_2, N_2). Additionally, hybrid nanoparticles that combine materials from different groups are being explored to leverage the benefits of multiple types of nanoparticles simultaneously. The design and properties of these nanoparticles are crucial for optimizing photoacoustic imaging. For instance, metal nanoparticles can exhibit surface plasmon resonance (SPR), which can be tuned to match the near-infrared (NIR) spectrum through modifications in shape and surface characteristics. Carbon nanomaterials are noted for their low heat capacity and exceptional light absorption properties. Nanoparticle structures vary from simple configurations with surface labeling to more complex core-shell and vector structures, each tailored for specific imaging needs. Applications of photoacoustic imaging nanoparticles are diverse, extending beyond tumor imaging to include cell tracking, lymph node imaging, vascular imaging, early detection of osteoarthritis, and monitoring of blood-brain barrier function. The versatility of these nanoparticles enables targeted and detailed imaging across various biological contexts, highlighting their potential to advance diagnostic and therapeutic approaches in biomedicine [61, 66]. Fig. (**5**) reflects the multifunctional nanoplatforms for medical imaging.

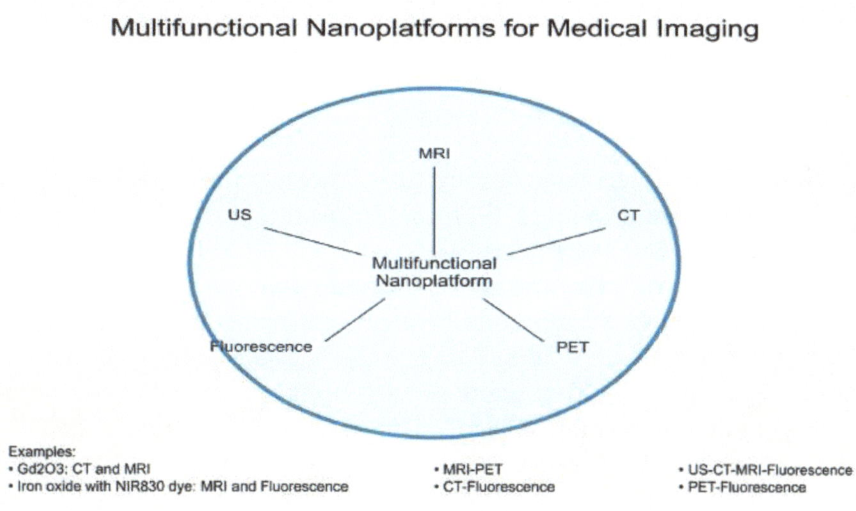

Fig. (5). Multifunctional nanoplatforms for medical imaging.

NANOPARTICLE-ENHANCED PET/SPECT IMAGING: A NEW ERA IN DIAGNOSTIC IMAGING

Pet/Spect Imaging: Strengths and Constraints

Positron Emission Tomography (PET) and Single Photon Emission Computed Tomography (SPECT) are advanced nuclear imaging techniques widely employed in both clinical and research environments due to their superior tissue penetration, enhanced sensitivity, and ability to provide real-time quantitative data. PET is particularly notable for its capacity to not only deliver anatomical insights but also monitor molecular-level biological processes by tracking radionuclides, making it indispensable in fields such as oncology and neurology [67]. SPECT, though similar to PET, utilizes different radiopharmaceuticals and detection mechanisms, while maintaining the ability to detect biochemical changes before they become anatomically apparent, which is critical for early disease detection and patient monitoring. Despite their considerable advantages, both PET and SPECT have limitations, including high operational costs stemming from the infrastructure needed for radioisotope production and the specialized imaging systems required [68]. Furthermore, the use of radioactive tracers raises concerns about radiation exposure, necessitating rigorous safety protocols and adherence to dosimetry guidelines to protect both patients and healthcare workers.

Structural and Compositional Innovations

The development of nanoparticle-based tracers for Positron Emission Tomography (PET) and Single Photon Emission Computed Tomography (SPECT) imaging requires the selection of radionuclides with extended half-lives. This necessity arises from the time needed for nanoparticle synthesis and cellular uptake. In PET imaging, the commonly used radionuclide Fluorine-18, with a half-life of 109.8 minutes, is often inadequate for nanoparticle-based systems, prompting the need for radionuclides with longer half-lives. Examples of radionuclides suitable for nanoparticle PET tracers include Copper-64 (half-life of 12.7 hours), Indium-111 (half-life of 2.8 days), and Iodine-124 (half-life of 4.2 days). The structural design of nanoparticles for PET is similar to that used in other imaging techniques, with the primary difference being the incorporation of radionuclide tracers. In SPECT imaging, radionuclides with longer half-lives are typically used. Technetium-99m, with a half-life of 6 hours, is the most widely used radionuclide in clinical SPECT applications. Nanoparticles designed for SPECT imaging share a similar architecture to those used in PET. A prominent example in SPECT is the use of Technetium-99m-labeled macro-aggregated albumin particles, which are widely applied in the measurement of tumor volumes and pulmonary shunt fractions in liver studies. Other radionuclides used in nanoparticle formulations for SPECT include Indium-111, Iodine-123, 125, and 131, as well as Gold-198 and 199. These radionuclides, with varying half-lives and emission properties, enable the customization of imaging strategies to match the biological distribution and diagnostic needs of nanoparticle-based tracers [69, 70].

Nanoparticles: Innovations in Biomedical Imaging

The development of nanoparticle-based tracers for Positron Emission Tomography (PET) and Single Photon Emission Computed Tomography (SPECT) imaging requires the selection of radionuclides with extended half-lives. This necessity arises from the time needed for nanoparticle synthesis and cellular uptake. In PET imaging, the commonly used radionuclide Fluorine-18, with a half-life of 109.8 minutes, is often inadequate for nanoparticle-based systems, prompting the need for radionuclides with longer half-lives. Examples of radionuclides suitable for nanoparticle PET tracers include Copper-64 (half-life of 12.7 hours), Indium-111 (half-life of 2.8 days), and Iodine-124 (half-life of 4.2 days), as shown in Table **4**. The structural design of nanoparticles for PET is similar to that used in other imaging techniques, with the primary difference being the incorporation of radionuclide tracers. In SPECT imaging, radionuclides with longer half-lives are typically used. Technetium-99m, with a half-life of 6 hours, is the most widely used radionuclide in clinical SPECT applications. Nanoparticles designed for SPECT imaging share a similar architecture to those used in PET. A prominent example in SPECT is the use of Technetium-99m-labeled macro-aggregated albumin particles, which are widely applied in the measurement of tumor volumes and pulmonary shunt fractions in liver studies. Other radionuclides used in nanoparticle formulations for SPECT include Indium-111, Iodine-123, 125, and 131, as well as Gold-198 and 199. These radionuclides, with varying half-lives and emission properties, enable the customization of imaging strategies to match the biological distribution and diagnostic needs of nanoparticle-based tracers [69, 70].

Table 4. Highlights a variety of nanoparticle formulations.

S.No.	Nanoparticle Composition	Imaging Agent	Size (nm)	Biomedical Application
1	Gold nanoparticles	T2-weighted	30-50	Tumor imaging
2	Iron oxide nanoparticles	T1-weighted	10-100	Lymph node mapping
3	Liposomes	T1-weighted	100-200	Drug delivery tracking
4	Quantum dots	Dual T1/T2	02-Oct	Multimodal imaging
5	Dendrimers	19F-MRI	May-15	Tumor angiogenesis

(Table 4) cont.....

S.No.	Nanoparticle Composition	Imaging Agent	Size (nm)	Biomedical Application
6	Silica nanoparticles	T2-weighted	50-200	Sentinel lymph node detection
7	Carbon nanotubes	T2-weighted	1-2 (diameter)	Stem cell tracking
8	Polymeric micelles	T1-weighted	20-100	Tumor targeting

PET/SPECT imaging: Notable Nanoparticle Applications

In the field of nuclear medicine, the integration of radiopharmaceutical nuclides into nanoparticle-based imaging platforms represents a significant advancement. A major challenge has been the detachment of radiolabeled probes from nanoparticle carriers, which can undermine imaging stability and accuracy. To address this issue, a novel approach has been developed involving the direct conjugation of radionuclides to stable nanoparticles, particularly for targeted Positron Emission Tomography (PET) imaging [75]. For instance, poly(4-vinylphenol) (PVPh) nanoparticles have been engineered to incorporate iodine-124 directly within their structure, and their surfaces are functionalized with antibodies targeting endothelial cells [76]. This innovative design enhances the stability of the imaging platform by minimizing radionuclide detachment, thus ensuring prolonged and reliable imaging. The targeted antibodies enable precise localization to endothelial cells, improving both the sensitivity and accuracy of imaging. Additionally, the stable incorporation of iodine-124 allows for extended image tracking, making this platform suitable for longitudinal studies and monitoring biological processes. This approach demonstrates considerable promise for clinical applications, as it combines stable radionuclide integration with molecular targeting strategies, highlighting the interdisciplinary efforts to advance diagnostic imaging technologies in nuclear medicine [77, 78]. Fig. (**6**) depicts the size comparison of imaging contrast agents.

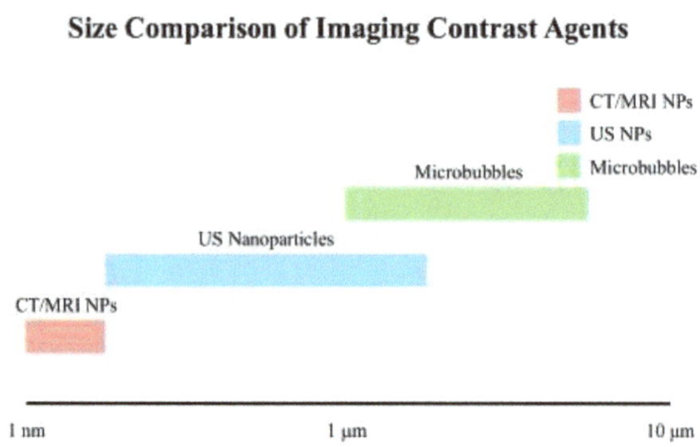

Fig. (6). The size comparison of imaging contrast agents.

NANOPARTICLE-ENHANCED ULTRASOUND IMAGING: A NEW ERA IN DIAGNOSTIC IMAGING

Ultrsound Imaging: Strengths and Constraints

Ultrasound (US) imaging is a widely employed diagnostic tool known for its portability, non-invasive nature, high spatial resolution, cost-effectiveness, and real-time imaging capabilities. To enhance its diagnostic performance, contrast agents have been developed to improve the visibility of tissues and blood vessels. Currently, most commercial ultrasound contrast agents are based on microbubbles, which range from 1 to 8 μm in diameter. These microbubbles enhance the acoustic signal by amplifying the differences between normal and targeted tissues. However, microbubble-based agents face certain limitations: their larger size confines them to providing blood pool contrast, impedes cellular uptake, limits their circulation time, and may compromise their stability *in vivo* [79]. To overcome these drawbacks, researchers are exploring nanoparticle-based contrast agents. Unlike microbubbles, nanoparticles must be sufficiently large to generate adequate acoustic reflections for effective imaging. Typically, ultrasound contrast nanoparticles range from hundreds to thousands of nanometers in size. This size range offers potential benefits, such as improved stability and extended circulation time, while also presenting challenges in maintaining the necessary acoustic properties for optimal imaging performance [80].

Structural and Compositional Innovations

Ultrasound (US) contrast nanoparticles, typically around 200 nanometers in size, offer several key benefits for medical imaging. Their small size enables functionalization with ligands for targeted molecular binding, facilitates cellular uptake, and allows for passive accumulation in tumors through the Enhanced Permeability and Retention (EPR) effect. For instance, folate-coated silica nanoparticles loaded with perfluoropentane have been used for prostate cancer imaging. There are three main types of ultrasound contrast nanoparticles: gas-based, solid-based, and liquid-based. Gas-based nanoparticles, such as those containing nitrogen or perfluorocarbon, generate strong acoustic reflections due to their gas core and are commonly used in imaging. Solid-based nanoparticles, like mesoporous silica structures, produce high scattering acoustic signals and accumulate in target lesions *via* cellular uptake or specific receptor binding. Liquid-based nanoparticles, such as those containing perfluorooctyl bromide (PFOB), create contrast through differences in the speed of sound transmission and acoustic reflection from their liquid core. Each type offers unique properties and applications, contributing to advancements in ultrasound imaging by improving diagnostic accuracy and imaging capabilities [81, 82]. Fig. (7) shows the different types of ultrasound contrast nanoparticles.

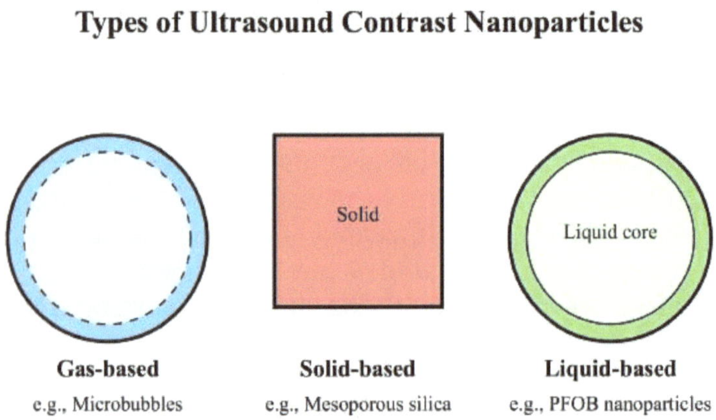

Fig. (7). The different types of ultrasound contrast nanoparticles.

Nanoparticles: Innovations in Biomedical Imaging

Perfluoropentane-loaded polymeric micelles, with sizes of 300-400 nm, serve in tumor detection and are in the *in vivo* experimental phase. Gold nanorods, solid-based and ranging from 50-100 nm, are applied in photothermal therapy and are currently tested *in vivo*. Chitosan-based nanobubbles, with dimensions of 200-500 nm, are used for gene delivery and are in vitro [83]. Mesoporous silica nanoparticles, sized 100-150 nm and used for theranostics, are also *in vivo*. PFOB-loaded PLGA nanoparticles, liquid-based and measuring 150-250 nm, are utilized for inflammation imaging and are undergoing *in vivo* studies [84]. Perfluorocarbon nanodroplets, with a size range of 200-400 nm and used for vascular imaging, are also *in vivo*. Table **4** shows these diverse agents span organic and inorganic materials, with sizes generally ranging from 100-500 nm, significantly smaller than traditional microbubble agents [85]. Their varied applications, from diagnostic imaging to therapeutic uses, and their progression to *in vivo* studies indicate their potential for clinical translation, as shown in Table **5**. The development of these nanoparticle-based contrast agents marks a notable leap forward in ultrasound imaging, integrating the advantages of nanotechnology with established imaging techniques to improve diagnostic accuracy and enable targeted therapies.

Table 5. Overview of various applications of nanoparticles in biomedical imaging.

S.No.	Nanoparticle Composition	Classification	Size (nm)	Biomedical Application
1	PFOB-loaded liposomes	Liquid-based	200-300	Tumor imaging
2	Silica nanoparticles	Solid-based	100-200	Drug delivery
3	Perfluoropentane-loaded polymeric micelles	Gas-based	300-400	Tumor detection
4	Gold nanorods	Solid-based	50-100	Photothermal therapy
5	Chitosan-based nanobubbles	Gas-based	200-500	Gene delivery
6	Mesoporous silica nanoparticles	Solid-based	100-150	Theranostics
7	PFOB-loaded PLGA nanoparticles	Liquid-based	150-250	Inflammation imaging
8	Perfluorocarbon nanodroplets	Liquid/Gas-based	200-400	Vascular imaging

Ultrasound Imaging: Notable Nanoparticle Applications

The incorporation of gaseous agents into nanoparticulate cores for ultrasound contrast enhancement can be achieved through three distinct methodologies, each offering specific advantages for diverse applications. The first method involves the direct encapsulation of naturally occurring gases—such as nitrogen, atmospheric air, perfluorocarbons, or sulfur hexafluoride—within the nanoparticle core. This approach leverages the intrinsic acoustic properties of these gases to generate contrast. The second strategy utilizes a phase-transition mechanism, where a liquid core vaporizes upon exposure to ultrasound waves [80]. This technique is especially effective when nanoparticles are targeted to accumulate in specific tissues through active or passive targeting. Upon ultrasound stimulation, the liquid core rapidly transitions to a gaseous state, resulting in enhanced acoustic reflection. An exemplary implementation of this method is seen in AuNP-coated, perfluorohexane-encapsulated, PEGylated mesoporous silica nanocapsules [81].

The third method involves in situ gas generation through chemical reactions, typically with carbonate-based compounds. In this approach, insoluble carbonate species are encapsulated within nanoparticles that accumulate in tumors *via* the enhanced permeability and retention (EPR) effect or specific binding mechanisms [82]. The acidic microenvironment of tumor tissues facilitates the reaction of the carbonate core, producing carbon dioxide gas and generating robust acoustic contrast. Notable examples include poly(cholesteryl carbonate γ-butyrolactone-propylene oxide) nanoparticles for tumor imaging and ammonium bicarbonate, which has also demonstrated efficacy in acoustic imaging applications. These gas-loading strategies offer a spectrum of options for tailoring nanoparticle-based ultrasound contrast agents to specific diagnostic and therapeutic requirements, enhancing visualization and enabling potential therapeutic interventions in various biomedical contexts.

CONCLUSION

Our primary focus is on examining the structural characteristics of nanoparticles and their relevant applications in biomedical imaging. This includes their use in cancer detection and imaging of other diseases. For each imaging modality, we will explore frequently researched examples to illustrate the potential and current state of nanoparticle-based imaging techniques. Currently, approximately 50 nanodrugs have received FDA approval. However, the transition of nanotechnology to clinical practice faces delays primarily due to concerns regarding biodistribution and safety. The way nanoparticles distribute throughout biological systems depends on multiple characteristics. One of the most significant determinants is their dimensional properties, but other attributes like geometric configuration, electrical properties of the surface, and surface chemical modifications also substantially impact their distribution patterns. Additionally, there remain significant knowledge gaps regarding the potential toxic effects that may arise when organisms process and break down these nanoparticles. Certain elements within nanoparticles may potentially harm normal cells, depending on their inherent physicochemical properties.

For instance, free-state rare earth elements from quantum dots have been shown to be harmful. While encapsulation and surface modifications have been explored as strategies to mitigate toxic effects, further research is necessary to gain a comprehensive understanding of nanoparticle behavior in biological systems. The development of safe and effective nanomaterials for medical imaging and drug delivery requires ongoing investigation into their long-term effects and interactions within the body. As the field progresses, addressing these concerns will be crucial for the successful translation of nanoparticle-based technologies into clinical practice. An exciting potential application for nanoparticle imaging agents is the integration of diagnostic and therapeutic capabilities within a single nanoplatform, a concept known as theranostics. Nanoparticles, serving as delivery vehicles, have shown great promise in encapsulating various therapeutic agents for targeted delivery during biomedical imaging procedures. The integration of diverse therapeutic modalities into diagnostic-therapeutic nanoplatforms has emerged as a key focus of contemporary research. These comprehensive treatment approaches encompass chemotherapeutic agents, photothermal interventions, radiation-based treatments, photodynamic methods, immune system modulation, and genetic manipulation techniques. Cancer research has been the primary field where this multi-modal therapeutic strategy has found extensive application. For instance, ball-in-ball ZrO_2 nanoparticles have been developed to provide CT imaging capabilities while simultaneously enabling synergistic microwave ablation and chemotherapy when accumulated in tumors. Despite the potential of theranostic nanoparticles, none have yet received FDA approval.

This is partly due to the significant variability in the design and synthesis of these agents. Substantial research efforts will be required in the future to advance this field and address the challenges associated with developing safe and effective theranostic nanoplatforms. As the field of nanomedicine progresses, the development of multifunctional nanoparticles that can simultaneously diagnose and treat diseases represents a promising avenue for improving patient care and treatment outcomes. However, overcoming regulatory hurdles and ensuring consistent manufacturing processes will be crucial for the clinical translation of these advanced nanoplatforms. To enhance the clinical translation of nanoparticle research, it is crucial to gain a comprehensive understanding of all nanoparticle characteristics, including composition and formulation, shape and surface charge, hydrodynamic diameter, solubility and stability, administration route, distribution, metabolism, and clearance, and potential toxic effects. The development of novel therapeutic agents requires extensive validation through comprehensive clinical trials to establish both their safety profile and therapeutic effectiveness.

The journey from initial discovery to clinical implementation through FDA approval typically spans approximately twelve years, with safety assessments remaining paramount throughout this process. Despite considerable advances in nano-contrast agent development, the vast majority remain at the experimental phase, with human trials being notably limited. The advancement of this field necessitates focused research efforts in several key areas: enhancing selective targeting capabilities, reducing potential toxic effects, strengthening clinical relevance through appropriate preclinical models, and developing a thorough understanding of human-specific pharmacological behaviors. The successful translation of nanoparticle-based technologies into clinical applications demands an interdisciplinary approach, bringing together expertise from chemistry, pharmacy, biology, physics, engineering, and medicine. While significant obstacles exist, nanoparticle-based imaging agents possess transformative potential across various clinical applications. The path forward requires careful attention to safety validation, performance optimization, and adherence to regulatory frameworks to fully realize the clinical potential of these innovative imaging tools.

REFERENCES

[1] G.P. Young, and B.H. Demediuk, "The genetics, epidemiology, and early detection of gastrointestinal cancers", *Curr. Opin. Oncol.,* vol. 4, no. 4, pp. 728-735, 1992. [http://dx.doi.org/10.1097/00001622-199208000-00019] [PMID: 1511028]

[2] T.J. Hoerger, D.U. Ekwueme, J.W. Miller, V. Uzunangelov, I.J. Hall, J. Segel, J. Royalty, J.G. Gardner, J.L. Smith, and C. Li, "Estimated effects of the national breast and cervical cancer early detection program on breast cancer mortality", *Am. J. Prev. Med.,* vol. 40, no. 4, pp. 397-404, 2011. [http://dx.doi.org/10.1016/j.amepre.2010.12.017] [PMID: 21406272]

[3] X. Ou, X. Chen, X. Xu, L. Xie, X. Chen, Z. Hong, and H. Bai, *Recent development in x-ray imaging technology: Future and challenges.* Research, 2021.

[4] R. Kumar, R.K. Ratnesh, J. Singh, R. Chandra, G. Singh, and V. Vishnoi, "Recent prospects of medical imaging and sensing technologies based on electrical impedance data acquisition system", *J. Electrochem. Soc.,* vol. 170, no. 11, p. 117507, 2023. [http://dx.doi.org/10.1149/1945-7111/ad050f]

[5] K.E. Tipirneni, E.L. Rosenthal, L.S. Moore, A.D. Haskins, N. Udayakumar, A.H. Jani, W.R. Carroll, A.B. Morlandt, M. Bogyo, J. Rao, and J.M. Warram, "Fluorescence imaging for cancer screening and surveillance", *Mol. Imaging Biol.,* vol. 19, no. 5, pp. 645-655, 2017. [http://dx.doi.org/10.1007/s11307-017-1050-5] [PMID: 28155079]

[6] A.S. Panayides, A. Amini, N.D. Filipovic, A. Sharma, S.A. Tsaftaris, A. Young, D. Foran, N. Do, S. Golemati, T. Kurc, K. Huang, K.S. Nikita, B.P. Veasey, M. Zervakis, J.H. Saltz, and C.S. Pattichis, "AI in medical imaging informatics: current challenges and future directions", *IEEE J. Biomed. Health Inform.,* vol. 24, no. 7, pp. 1837-1857, 2020. [http://dx.doi.org/10.1109/JBHI.2020.2991043] [PMID: 32609615]

[7] Y.E. Chung, and K.W. Kim, "Contrast-enhanced ultrasonography: advance and current status in abdominal imaging", *Ultrasonography,* vol. 34, no. 1, pp. 3-18, 2015. [http://dx.doi.org/10.14366/usg.14034] [PMID: 25342120]

[8] S. Siddique, and J.C.L. Chow, "Application of nanomaterials in biomedical imaging and cancer therapy", *Nanomaterials (Basel),* vol. 10, no. 9, p. 1700, 2020. [http://dx.doi.org/10.3390/nano10091700] [PMID: 32872399]

[9] Y. Du, X. Liu, Q. Liang, X.J. Liang, and J. Tian, "Optimization and design of magnetic ferrite nanoparticles with uniform tumor distribution for highly sensitive MRI/MPI performance and improved magnetic hyperthermia therapy", *Nano Lett.,* vol. 19, no. 6, pp. 3618-3626, 2019. [http://dx.doi.org/10.1021/acs.nanolett.9b00630] [PMID: 31074627]

[10] Andrea Kunzmann, Britta Andersson, Tina Thurnherr, Harald Krug, Annika Scheynius, and Bengt Fadeel, "Toxicology of engineered nanomaterials: focus on biocompatibility, biodistribution and biodegradation", *Biochimica et Biophysica Acta (BBA)-general subjects 1810,* vol. 3, no. 2011, pp. 361-373, 2011. [http://dx.doi.org/10.1016/j.bbagen.2010.04.007]

[11] C. Zhou, M. Long, Y. Qin, X. Sun, and J. Zheng, "Luminescent gold nanoparticles with efficient renal clearance", *Angew. Chem. Int. Ed.,* vol. 50, no. 14, pp. 3168-3172, 2011. [http://dx.doi.org/10.1002/anie.201007321] [PMID: 21374769]

[12] M. Yu, and J. Zheng, "Clearance pathways and tumor targeting of imaging nanoparticles", *ACS Nano,* vol. 9, no. 7, pp. 6655-6674, 2015. [http://dx.doi.org/10.1021/acsnano.5b01320] [PMID: 26149184]

[13] K. Saha, M. Rahimi, M. Yazdani, S.T. Kim, D.F. Moyano, S. Hou, R. Das, R. Mout, F. Rezaee, M. Mahmoudi, and V.M. Rotello, "Regulation of macrophage recognition through the interplay of nanoparticle surface functionality and protein corona", *ACS Nano,* vol. 10, no. 4, pp. 4421-4430, 2016. [http://dx.doi.org/10.1021/acsnano.6b00053] [PMID: 27040442]

[14] H.S. Choi, and J.V. Frangioni, "Nanoparticles for biomedical imaging: fundamentals of clinical translation", *Mol. Imaging,* vol. 9, no. 6, p. 7290.2010.00031, 2010. [http://dx.doi.org/10.2310/7290.2010.00031] [PMID: 21084027]

[15] L. Wang, Y. Wang, and Z. Li, "Nanoparticle-based tumor theranostics with molecular imaging",

Current Pharmaceutical Biotechnology, vol. 4, no. 7, pp. 3689-3696, 2013.

[16] G. Huang, C. Zhang, S. Li, C. Khemtong, S.G. Yang, R. Tian, J.D. Minna, K.C. Brown, and J. Gao, "A novel strategy for surface modification of superparamagnetic iron oxide nanoparticles for lung cancer imaging", *J. Mater. Chem.,* vol. 19, no. 35, pp. 6367-6372, 2009. [http://dx.doi.org/10.1039/b902358e] [PMID: 20505790]

[17] E. Blanco, H. Shen, and M. Ferrari, "Principles of nanoparticle design for overcoming biological barriers to drug delivery", *Nat. Biotechnol.,* vol. 33, no. 9, pp. 941-951, 2015. [http://dx.doi.org/10.1038/nbt.3330] [PMID: 26348965]

[18] X. Liu, H. Zhang, T. Zhang, Y. Wang, W. Jiao, X. Lu, X. Gao, M. Xie, Q. Shan, N. Wen, C. Liu, W.S.V. Lee, and H. Fan, "Magnetic nanomaterials-mediated cancer diagnosis and therapy", *Prog. Biomed. Eng. (Bristol),* vol. 4, no. 1, p. 012005, 2022. [http://dx.doi.org/10.1088/2516-1091/ac3111] [PMID: 41287228]

[19] M. Kumar Yadav, R. Kumar, R. Kumar Ratnesh, J. Singh, R. Chandra, A. Kumar, V. Vishnoi, G. Singh, and A. Kumar Singh, "Revolutionizing technology with spintronics: Devices and their transformative applications", *Mater. Sci. Eng. B,* vol. 303, p. 117293, 2024. [http://dx.doi.org/10.1016/j.mseb.2024.117293]

[20] A.S. Barnard, "Diamond standard in diagnostics: nanodiamond biolabels make their mark", *Analyst (Lond.),* vol. 134, no. 9, pp. 1751-1764, 2009. [http://dx.doi.org/10.1039/b908532g] [PMID: 19684895]

[21] H.B. Na, I.C. Song, and T. Hyeon, "Inorganic nanoparticles for MRI contrast agents", *Adv. Mater.,* vol. 21, no. 21, pp. 2133-2148, 2009. [http://dx.doi.org/10.1002/adma.200802366]

[22] A. Foucault-Collet, K.A. Gogick, K.A. White, S. Villette, A. Pallier, G. Collet, C. Kieda, T. Li, S.J. Geib, N.L. Rosi, and S. Petoud, "Lanthanide near infrared imaging in living cells with Yb 3+ nano metal organic frameworks", *Proc. Natl. Acad. Sci. USA,* vol. 110, no. 43, pp. 17199-17204, 2013. [http://dx.doi.org/10.1073/pnas.1305910110] [PMID: 24108356]

[23] E.S. Hwang, C. Cao, S. Hong, H.J. Jung, C.Y. Cha, J.B. Choi, Y.J. Kim, and S. Baik, "The DNA hybridization assay using single-walled carbon nanotubes as ultrasensitive, long-term optical labels", *Nanotechnology,* vol. 17, no. 14, pp. 3442-3445, 2006. [http://dx.doi.org/10.1088/0957-4484/17/14/016] [PMID: 19661588]

[24] R. Kumar, R.K. Ratnesh, J. Singh, A. Kumar, and R. Chandra, "IoT-driven experimental framework for advancing electrical impedance tomography", *ECS J. Solid State Sci. Technol.,* vol. 13, no. 2, p. 027002, 2024. [http://dx.doi.org/10.1149/2162-8777/ad2331]

[25] P.K. Ramesh, R. Yadav, and H. Wong, "Metallic nanoparticles for CT contrast: A comparative analysis of gold and tantalum-based agents", *IEEE Access,* vol. 9, pp. 13456-13465, 2021. [http://dx.doi.org/10.1109/ACCESS.2021.3057609]

[26] J.C. Nguyen, A.M. Castillo, and Y. Park, "Advances in core-shell nanoparticles for CT imaging", *IEEE Trans. Biomed. Eng.,* vol. 68, no. 4, pp. 1283-1290, 2021. [http://dx.doi.org/10.1109/TBME.2021.3048902]

[27] Z. Zhang, T. Wu, and L. Huang, "Applications of gold nanoparticles in CT contrast agents for cancer imaging", *IEEE Rev. Biomed. Eng.,* vol. 14, pp. 352-367, 2021. [http://dx.doi.org/10.1109/RBME.2021.3061234]

[28] W.Y. Fan, X.H. Huang, and B. Xu, "Gadolinium-poly(acrylic acid) hybrid nanoparticles for dual-mode CT/MRI imaging", *IEEE Trans. Nanobiosci.,* vol. 19, no. 7, pp. 583-589, 2020. [http://dx.doi.org/10.1109/TNB.2020.3001253]

[29] X. Zhang, "Advancements in nanoparticle-based contrast agents for computed tomography imaging of tumors", *IEEE Nanotechnol. Mag.,* vol. 15, no. 3, pp. 38-45, 2021.

[http://dx.doi.org/10.1109/MNANO.2021.3082197]

[30] A. Jindal, and A. Mainuddin, Kumar, R. K. Ratnesh, J. Singh, "Nanotechnology Driven Lipid and Metalloid Based Formulations Targeting Blood-Brain Barrier (3B) for Brain Tumor", *Indian J. Microbiol.,* 2024.

[31] K. Shi, S. Liu, X. Zhou, H. Lv, and L. Wang, "Development of iodine-encapsulated nanoparticles for enhanced CT imaging of blood vessels", *IEEE Trans. Biomed. Eng.,* vol. 68, no. 5, pp. 1472-1480, 2021.
[http://dx.doi.org/10.1109/TBME.2020.3048537]

[32] Z. Guo, Y. Song, and J. Liu, "Gold nanoparticle-based CT contrast agents for cancer diagnosis and therapy", *IEEE Access,* vol. 9, pp. 20348-20357, 2021.
[http://dx.doi.org/10.1109/ACCESS.2021.3050845]

[33] M.R. García, F. Rodríguez, and A. López, "Emerging nanotechnology applications in computed tomography: Gold and iodine nanoparticles", *IEEE Nanotechnol. Mag.,* vol. 15, no. 3, pp. 29-35, 2021.
[http://dx.doi.org/10.1109/MNANO.2021.3050170]

[34] J. Chen, X. Chen, and Q. Ren, "Near-infrared fluorescence imaging with nanoparticle-based contrast agents", *IEEE Trans. Biomed. Eng.,* vol. 69, no. 1, pp. 125-134, 2022.
[http://dx.doi.org/10.1109/TBME.2021.3084567] [PMID: 34185636]

[35] M.Y. Park, J.W. Kim, and C.H. Lee, "Nanoparticle-based NIR fluorescence imaging for cancer detection", *IEEE Access,* vol. 9, pp. 54789-54797, 2021.
[http://dx.doi.org/10.1109/ACCESS.2021.3070059]

[36] R. Zhang, X. Fang, and J. Wang, "FRET and quenching mechanisms in nanoparticle-based fluorescence imaging", *IEEE Trans. Nanotechnol.,* vol. 21, pp. 413-420, 2022.
[http://dx.doi.org/10.1109/TNANO.2022.3174421]

[37] R Kumar, and S Tripathi, "A Novel GUI-Based Image Reconstruction Algorithm of EIT Imaging Technique", *International Journal of Cognitive Informatics and Natural Intelligence (IJCINI) 15,* vol. 3, no. 2021, pp. 31-46, 2021.
[http://dx.doi.org/10.4018/IJCINI.20210701.oa3]

[38] Z. Li, Y. Liu, and L. Zhang, "Upconversion nanoparticles for NIR fluorescence imaging: Challenges and advances", *IEEE Trans. Biomed. Eng.,* vol. 69, no. 3, pp. 830-837, 2022.
[http://dx.doi.org/10.1109/TBME.2021.3105768]

[39] H. Kim, J.J. Sung, and S. Yoon, "Nanoparticle-based fluorescence imaging for tumor diagnosis and real-time therapeutic monitoring", *Biosens. Bioelectron.,* vol. 190, p. 113353, 2021.

[40] Y. Zhang, Q. Zhao, and J. Wang, "Advances in nanoparticle-based fluorescent probes for *in vivo* bioimaging", *ACS Nano,* vol. 15, no. 9, pp. 13700-13714, 2022.

[41] M. Li, X. Shen, and Y. Chen, "Nanoparticle-mediated fluorescence imaging: A bridge between molecular biology and clinical applications", *J. Biomed. Nanotechnol.,* vol. 19, no. 1, pp. 45-63, 2023.

[42] L. Chen, R. Wei, J. Dong, and Y. Liu, "Quantum dots for bioimaging: Recent advances and challenges", *Adv. Mater.,* vol. 35, no. 2, p. 1203467, 2023.

[43] X. Wang, M. Zheng, and J. Zhang, "Fluorescent quantum dots: Toxicity, biocompatibility, and applications in biomedical imaging", *J. Nanobiotechnology,* vol. 21, no. 5, pp. 65-78, 2023.

[44] C. Liu, Y. Li, and X. Zhou, "Recent progress in the application of gold nanoparticles for biomedical fluorescence imaging", *ACS Nano,* vol. 17, no. 4, pp. 12245-12258, 2023.

[45] S. Xue, Y. Zhou, and G. Yang, "Advances in the development of MRI contrast agents for enhanced diagnostic imaging", *J. Magn. Reson. Imaging,* vol. 59, no. 1, pp. 245-258, 2023.
[PMID: 37881827]

[46] L. Zhang, Y. Li, and M. Liu, "Recent progress in MRI contrast agents: Nanomaterials and molecular design", *Adv. Healthc. Mater.,* vol. 12, no. 5, p. 2201147, 2023.

[47] Y. Wang, L. Ma, and J. Zhang, "Recent advances in T1 and T2 MRI contrast agents for theranostic applications", *Adv. Funct. Mater.*, vol. 34, no. 2, p. 123040, 2023.

[48] M. Li, X. Song, and Y. Zhao, "Design and application of nanoparticle-based ^19F MRI contrast agents in biomedical imaging", *J. Nanomed. Res.*, vol. 18, no. 3, pp. 254-262, 2023.

[49] L. Zhang, J. Liu, and Z. Zhang, "Recent advances in nanoparticle-based MRI contrast agents for tumor imaging", *J. Magn. Reson. Imaging,* vol. 48, no. 1, pp. 45-62, 2023.
[PMID: 35993550]

[50] S. Xu, W. Li, and X. Chen, "Multifunctional nanoparticle MRI contrast agents for inflammation and cancer diagnostics", *Adv. Healthc. Mater.*, vol. 12, no. 3, p. 2200567, 2023.

[51] M.A. Green, J. Kim, and L. Park, "Advances in gadolinium nanoparticle MRI contrast agents: A review of recent developments and clinical implications", *J. Magn. Reson. Imaging,* vol. 50, no. 4, pp. 1234-1245, 2023.
[PMID: 36727433]

[52] P. Wang, S. Zhao, and X. Li, "Manganese-based nanoparticles as MRI contrast agents for cancer diagnosis and lymph node imaging", *J. Biomed. Nanotechnol.,* vol. 19, no. 2, pp. 445-458, 2023.

[53] Ramesh Kumar, Sharvan Kumar, and A. Sengupta, "An experimental analysis and validation of electrical impedance tomography technique for medical or industrial application", *Biomedical Engineering: Applications, Basis and Communications 31,* vol. 2, p. 1950010, 2019.
[http://dx.doi.org/10.4015/S1016237219500108]

[54] Y. Zhang, S. Li, and Z. Huang, "Recent progress in photoacoustic and PET/MRI dual-modality imaging with multifunctional nanoparticles", *J. Biomed. Nanotechnol.,* vol. 19, no. 4, pp. 873-890, 2023.

[55] X. Wang, Y. Zhang, and J. Li, "Recent advances in multimodal nanoplatforms for cancer imaging and therapy", *ACS Nano,* vol. 17, no. 3, pp. 2892-2907, 2023.

[56] L. Chen, Z. Liu, and H. Liu, "Multimodal imaging nanoplatforms for theranostic applications", *Nano Res.,* vol. 16, no. 5, pp. 1347-1360, 2023.

[57] M. Zhang, F. Tian, and D. Yang, "Multifunctional nanoparticles for MRI and optical imaging", *Adv. Mater.,* vol. 35, no. 14, pp. 3112-3125, 2023.

[58] T.M. Kelly, A.J. Robinson, and N.H. Patel, "Multimodal imaging nanoparticles for early cancer detection and treatment", *Theranostics,* vol. 13, no. 6, pp. 897-910, 2023.

[59] S.K. Gupta, M.R. Singh, and J.H. Lee, "Advancements in multimodal nanoparticle imaging for cardiovascular and neurological applications", *J. Control. Release,* vol. 355, pp. 450-463, 2023.

[60] Y. Zhang, J. Zhang, and L. Chen, "Recent developments in photoacoustic imaging with nanoparticles", *J. Biomed. Opt.,* vol. 28, no. 1, pp. 123-139, 2024.

[61] S. Kumar, M.R. Ali, and S.H. Park, "Nanoparticle-based contrast agents for enhanced photoacoustic imaging: A review", *Adv. Funct. Mater.,* vol. 34, no. 8, pp. 4596-4610, 2023.

[62] M. K., Yadav, R. Kumar, R.K. Ratnesh, J. Singh, R. Chandra, A. Kumar, & A.K. Singh, "Revolutionizing Technology with Spintronics: Devices and Their Transformative Applications". Mater. Sci. Eng.: B, 303, 117293, 2024.

[63] A.K. Singh, R. Kumar, H. Maity, P Singh & S. Singh, "Analog and linearity performance analysis of ferroelectric vertical tunnel field effect transistor with and without source pocket". *International Journal of Numerical Modelling: Electronic Networks, Devices and Fields*, 37(4), e3274, 2024.

[64] L. Wang, Y. Hu, and Y. Zhang, "Hybrid nanoparticles for multimodal photoacoustic imaging", *Nanomedicine,* vol. 47, pp. 134-145, 2023.

[65] J.S. Kim, Y.T. Lee, and K.M. Shin, "Innovations in photoacoustic imaging using carbon-based nanoparticles", *Mater. Today,* vol. 66, pp. 192-205, 2023.

[66] R.G. Burch, T.A. Thomas, and A.R. Williams, "Recent advancements in PET imaging technology and its clinical applications", *J. Nucl. Med.,* vol. 64, no. 7, pp. 1023-1032, 2024.

[67] R.K. Meena, "Electrical impedance tomography: A real-time medical imaging technique", In: *Handbook of research on advanced concepts in real-time image and video processing.* IGI Global, 2018, pp. 130-152.
[http://dx.doi.org/10.4018/978-1-5225-2848-7.ch006]

[68] X. Chen, H. Li, and J. Wang, "Development of advanced PET imaging tracers: Innovations and applications", *J. Nucl. Med.,* vol. 66, no. 4, pp. 658-669, 2024.

[69] M.A. Thompson, J.W. Johnson, and R.L. Edwards, "Recent advancements in SPECT imaging: Radionuclide selection and nanoparticle applications", *Med. Phys.,* vol. 51, no. 9, pp. 2301-2314, 2023.

[70] J. Smith, L. Zhang, and P. Lee, "Nanoparticles for PET and SPECT imaging: Mechanisms of accumulation and applications", *Adv. Drug Deliv. Rev.,* vol. 173, pp. 48-60, 2023.

[71] K. Patel, M. Williams, and R. Jones, "Clinical and research applications of nanoparticle-based imaging agents in PET and SPECT", *J. Nucl. Med.,* vol. 65, no. 9, pp. 1551-1562, 2024.

[72] L. Chen, S. Kumar, and Y. Lee, "Systematic evaluation of nanoparticle-based tracers in PET and SPECT imaging", *Med. Phys.,* vol. 51, no. 10, pp. 2411-2425, 2023.

[73] T. Anderson, H. Williams, and M. Smith, "Diverse nanoparticle formulations for molecular imaging: A comprehensive review", *Nanomedicine,* vol. 20, pp. 125-138, 2024.

[74] A.B. Goel, A.K. Sinha, and M. Shah, "Nanoparticle-based PET imaging agents: Current progress and future directions", *IEEE Trans. Biomed. Eng.,* vol. 67, no. 9, pp. 2356-2367, 2020.

[75] R.K. Jain, Y.S. Kang, J. Kim, and H. Park, "Radiolabeled nanoparticles for PET imaging: Design strategies and applications", *IEEE Access,* vol. 8, pp. 14267-14279, 2020.

[76] L. Chen, Z. Xu, T. Li, and J. Zhang, "Functionalized nanoparticles for targeted PET imaging: Advances and challenges", *IEEE Trans. Nanobiosci.,* vol. 19, no. 4, pp. 523-531, 2020.

[77] M. Zhao, S. Ren, and X. Xu, "Hybrid nanoparticles for enhanced molecular imaging and therapy", *IEEE Nanotechnol. Mag.,* vol. 15, no. 1, pp. 33-41, 2021.

[78] K.W. Ferrara, M.A. Borden, and H. Zhang, "Contrast agents for ultrasound imaging and ultrasound-mediated drug delivery", *Annu. Rev. Biomed. Eng.,* vol. 9, pp. 415-447, 2007.
[http://dx.doi.org/10.1146/annurev.bioeng.8.061505.095852] [PMID: 17651012]

[79] J.L. Gessner, and M.A. Dayton, "Advances in molecular imaging with ultrasound", *IEEE Trans. Ultrason. Ferroelectr. Freq. Control,* vol. 60, no. 7, pp. 1273-1282, 2013.

[80] S. Kooiman, H.J. Vos, M. Versluis, and N. de Jong, "Acoustic behavior of microbubbles and implications for contrast ultrasound imaging", *Adv. Drug Deliv. Rev.,* vol. 72, pp. 28-48, 2014.
[http://dx.doi.org/10.1016/j.addr.2014.03.003] [PMID: 24667643]

[81] R. Yin, J. Zhou, F. Xu, and Y. Wang, "Recent progress in application of multifunctional nanoparticles for enhanced ultrasound imaging", *IEEE Nanotechnol. Mag.,* vol. 13, no. 1, pp. 28-33, 2019.

[82] A.S.L. Rosa, J.S. Moraes, and P.S. Caldas, "Ultrasound contrast agents: Fundamentals, radiology, and nanotechnology", *J. Appl. Res. Technol.,* vol. 15, no. 4, pp. 394-403, 2017.
[http://dx.doi.org/10.1016/j.jart.2017.07.001]

[83] J.F.M. Goertz, R.C. Gessner, and M.L. O'Reilly, "Nanoparticle-based ultrasound contrast agents", *IEEE Trans. Ultrason. Ferroelectr. Freq. Control,* vol. 65, no. 8, pp. 1339-1352, 2018.
[http://dx.doi.org/10.1109/TUFFC.2018.2829423]

[84] T.R. Wilson, L.F. Carroll, and M.S. Robson, "The evolving role of nanotechnology in ultrasound imaging", *IEEE Eng. Med. Biol. Mag.,* vol. 36, no. 5, pp. 76-84, 2017.
[http://dx.doi.org/10.1109/MEMB.2017.2721976]

[85] Z. Li, Y. Wang, and L. Wang, "Applications of nanoparticles in molecular imaging and drug delivery", *IEEE Nanotechnol. Mag.,* vol. 12, no. 3, pp. 26-35, 2018. [http://dx.doi.org/10.1109/MNANO.2018.2865013]

CHAPTER 4

3d Local Descriptor-Based Abnormality Detection in Traffic Surveillance Videos

Gajendra Singh[1], **Ramesh Kumar**[2,*], **Vishal Vishnoi**[3], **Manoj Kumar**[4] and **Ashish Kumar Singh**[5]

[1] *School of Engineering and Sciences, G. D. Goenka University, Gurugram, India-122103*

[2] *Department of Electronics and Communication Engineering, Manipal University Jaipur, Jaipur, Rajasthan, India*

[3] *Department of Electronics and Communication Engineering, National Institute of Technology Sikkim, Sikkim, India*

[4] *School of Electrical and Electronics Engineering, Lovely Professional University, Phagwara, India*

[5] *Department of Electronics and Communication Engineering, Graphic Era (Deemed to be University), Dehradun, India*

Abstract: This research work is focused on detecting visual anomalies/abnormal events in video surveillance systems. In such situations, human attention may wander, and abnormal situations may be more challenging to note. So, to avoid such situations, an automated system is needed that can analyze such a huge amount of data and trigger alarms in abnormal events. Nowadays, the automation of surveillance systems is a major research area. This chapter presents a 3D local descriptor-based anomaly detection method for traffic surveillance videos, capable of extracting information about the appearance and motion of objects. Appearance information is extracted by 3D Histogram of Gradients (HOG), and motion information is extracted by 3D Histogram of Optical Flow Orientation (HOOF). Finally, these features are fed to a combined classifier for the detection of abnormality. Appearance information plays an important role when different types of objects are in the scene, like in traffic surveillance videos. This algorithm is tested on the YouTube video due to the unavailability of a publicly available standard traffic surveillance dataset.

Keywords: Abnormality detection, Histogram of gradients, Optical flow orientation, Surveillance systems, Traffic surveillance.

* **Corresponding author Ramesh Kumar:** Department of Electronics and Communication Engineering, Manipal University Jaipur, Jaipur, Rajasthan, India; E-mail: rameshkumarmeena@gmail.com

Ramesh Kumar, Anupma Gupta, Manish Singla, Arjun Puri, Ashish Kumar Singh & Vipan Kumar (Eds.)

INTRODUCTION

With the rapid development of video surveillance technologies in recent years, a greater number of smart video surveillance systems are being deployed in smart transport systems [1 - 4]. Traditional visual surveillance systems looking at roads or intersections have human intervention, which is not so effective. On the other hand, a smart transport system has very little human intervention. When something abnormal happens on roads, like road work, accidents, or jams, abnormal event detection systems detect the event and issue warning information [5 - 8]. So, automatic and timely detection of abnormal events will make our transportation system work efficiently and also reduce the workload of persons involved in the management of traffic [9 - 13].

PROPOSED METHOD

The block diagram of our proposed algorithm for the detection of abnormality in traffic surveillance systems is shown in Fig. (**1**). It is very important that extracted features are able to describe the scene properly. In this algorithm, the input video is described by a local feature descriptor, *viz.* 3D Histogram of Gradients (HOG) as proposed [1 - 3] and 3D Histogram of Optical Flow Orientation (HOOF). Both appearance and motion information are embedded in the feature vector. 3D HOG is used for extracting information about the appearance of different objects present in the scene. 3D HOOF feature extractors are based on optical flow, which provides information about the motion of objects. Finally, these extracted features are fed to the classifier for classification.

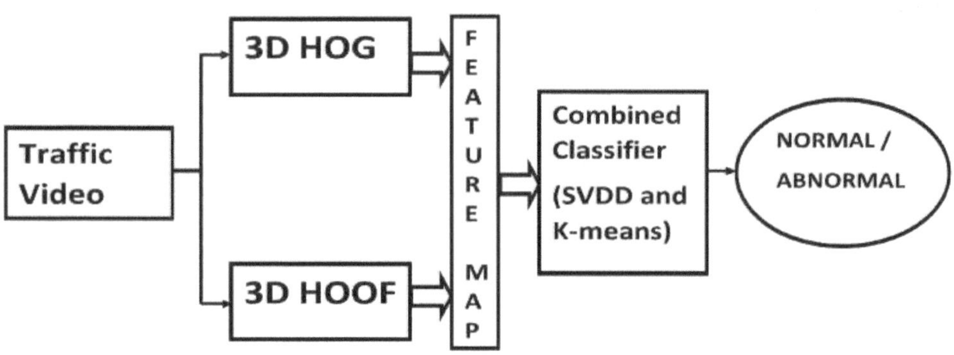

Fig. (1). Block diagram of our proposed algorithm.

Local Feature Extraction

In traffic surveillance video, the traffic scene is visually complex, with a large number of different types and sizes of cars, trucks, pedestrians, and objects (buildings, trees, *etc.*) that are unrelated to traffic. Fig. (**2**) shows the two scenes at different locations, both of which have a very large field of view.

Fig. (2). Images of traffic scenes.

Therefore, when using global image descriptors to describe such scenes, it becomes very challenging to accurately classify traffic scenes and identify individual objects within them. In a scene with a large field of view, an abnormal situation might occupy only a small portion, and the global descriptor may fail to capture it. For instance, if an accident occurs on the road, various events can happen simultaneously in different parts of the traffic scene. Some individuals might exhibit normal behavior, while others might attempt illegal U-turns, causing disruptions. Generally, accidents lead to traffic slowdowns in the affected lane and can also impact the flow in adjacent lanes [14 - 16].

To solve the above-mentioned problem, the feature descriptor needs to be calculated in local regions of the image or video. First, this algorithm divides the surveillance video into a number of video cubes. Then, a descriptor is calculated for each video cube, and the final scene is described by fusing the descriptor of each video cube [17 - 20].

Let A is a traffic video which is divided into N video cubes, also called cells, and expressed as follows:

$$A = C_1 \cup C_2 \cup \ldots \cup C_N \quad \textbf{(1)}$$

$C_k(1 \leq k \leq N)$ is k^{th} video cell of surveillance video A. The total number of video frames in each cell is kept the same. Further, each cell is sub-divided into a number of blocks as represented in Fig. (**3**).

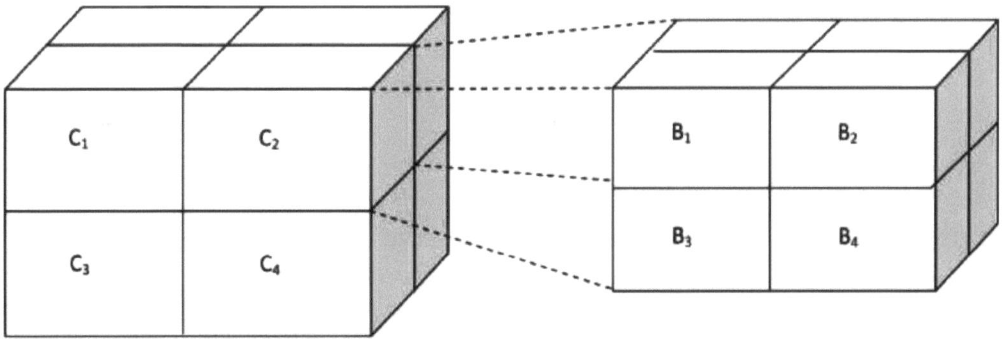

Fig. (3). Overview of descriptor computation.

3D Histogram of Oriented Gradients (3D-HOG)

To extract 3D-HOG [1], firstly, each video is divided into cells, and each cell is further divided into blocks. The process of computation of 3D-HOG is as follows:

Mean gradient vector computation: The Integral video concept is used for computing the mean gradient vector. This makes process memory efficient and flexible. This process can be considered a smoothing operation convolved with a box filter, replacing the local descriptor-based abnormality detection in traffic surveillance video, using a commonly used Gaussian filter. By using integral video, the mean 3D gradient is efficiently calculated for any choice of scale [4].

The integral video concept has been popularized for such applications. This type of representation has been used for the calculation of Haar features [21–23].

For a given video $A(x,y,t)$, its partial derivative along spatial (x, y) and temporal space (t) are given as $A_{\partial x}$, $A_{\partial y}$, $A_{\partial t}$. The integral video of the partial derivative ($A_{\partial x}$) can be calculated as proposed [1]:

$$iA_{\partial x} = \sum_{x' \leq x, y' \leq y, t' \leq t} A_{\partial x}(x', y', t') \quad \textbf{(2)}$$

Similarly, the integral videos of partial derivatives ($A_{\partial y}$, $A_{\partial t}$) are calculated.

For any 3D block of a video, the mean gradient vector can be given as

$$\bar{\mathbf{G}}_{\mathbf{B}} = \begin{pmatrix} \bar{G}_{B\,\partial x} \\ \bar{G}_{B\,\partial y} \\ \bar{G}_{B\,\partial t} \end{pmatrix}$$

$$\bar{G}_{B\,\partial x} = [iA_{\partial x}(x+w,y+h,t+l) - iA_{\partial x}(x,y+h,t+l) - iA_{\partial x}(x+w,y,t+l) + iA_{\partial x}(x,y,t+l) - iA_{\partial x}(x+w,y+h,t) + iA_{\partial x}(x,y+h,t) + iA_{\partial x}(x+w,y,t) - iA_{\partial x}(x,y,t)] \quad \textbf{(3)}$$

Similarly G ¯_B∂t and G ¯_B∂y are calculated.

Gradient orientation quantization: In the case of a static image (*i.e.*, 2D), a histogram of orientations having n number of bins can be thought of as a circle with a n sided polygon. Every side of a polygon is analogous to the bin of a histogram. While in the case of 3D space, a polygon will become a polyhedron. Regular polyhedra, having congruent faces, are called Platonic solids. There are only five types: the tetrahedron (4 sides), cube (6 sides), octahedron (8 sides), dodecahedron (12 sides), and icosahedron (20 sides), as shown in Fig. (**4**). In this work, the 20-sided icosahedron is used to quantize orientations in 3D because, as noted in a previous study [1 - 3], it exhibits a high degree of symmetry and provides a relatively uniform distribution of quantization levels across the unit sphere, leading to a more accurate and robust representation of 3D orientations.

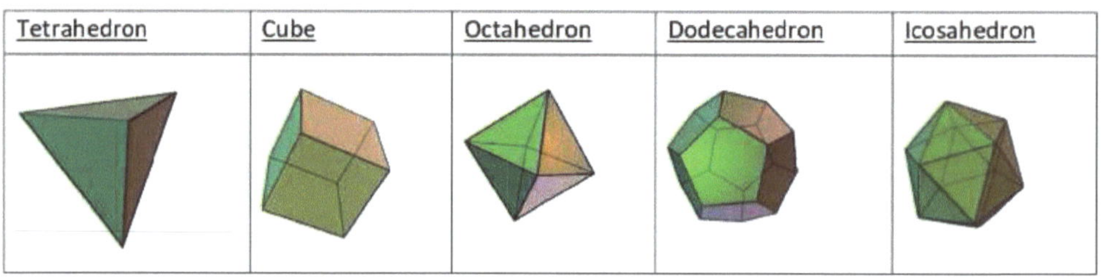

Fig. (4). Polyhedra having congruent faces.

For a regular n-sided polyhedron, suppose that in a 3D Euclidean coordinate system, the center of gravity is at the origin. To quantize the mean gradient vector G¯_B with respect to its orientation, G¯_B is projected on the axes going through the central positions of all faces of the polyhedron and the origin of the Euclidean coordinate system.

This is done by multiplying a matrix. Let *Cp* as a matrix of center positions of all faces and can be represented as:

$$C_p = \left[C_{p_1}, C_{p_2}, \dots, C_{p_n}\right]^T \quad \textbf{(4)}$$

Where $C_{pj} = (x_j, y_j, t_j)^T$

Suppose an icosahedronhas 20 faces and its 20 center points are as follows:

$$(\pm1, \pm1, \pm1)\left(0, \pm\frac{1}{\phi}, \pm\phi\right)\left(\pm\frac{1}{\phi}, \pm\phi, 0\right)\left(\pm\phi, 0, \pm\frac{1}{\phi}\right) \quad \textbf{(5)}$$

where ϕ is called the golden ratio and is equal to $\phi = \frac{1+\sqrt{5}}{2}$.

The projection of ϕ_B can be given as

$$\widehat{\mathbf{pb}_\mathbf{b}} = \left(\widehat{p_{b_1}, p_{b_2}}, \ldots, \widehat{p_{b_n}}\right)^T = \frac{C_p \cdot \overline{\mathbf{G}}_\mathbf{B}}{\|\overline{\mathbf{G}}_\mathbf{B}\|_2} \quad \textbf{(6)}$$

Each $\widehat{p_{b_j}}$ is normalized projection of G¯_B onto the axes going through the jth face center C_{pj}

$$\widehat{p_{b_j}} = \left\|C_{p_j}\right\|_2 \cdot \cos\angle\left(C_{p_j}, \overline{\mathbf{G}}_\mathbf{B}\right) = \|\overline{\mathbf{G}}_\mathbf{B}\|_2^{-1} \cdot \mathbf{C}_{\mathbf{p_j}}^\mathbf{T} \cdot \overline{\mathbf{G}}_\mathbf{B}. \quad \textbf{(7)}$$

The projected vector $\widehat{\mathbf{p}_\mathrm{b}}$ is thresholded because G¯_B should be voted only into the single bin where it is perfectly aligned with respective axes going through the face center and origin. The threshold can be obtained by comparing two adjacent axes C_{pj}, C_{pk} *i.e*, $th = C_{p\,j}^T.\ C_{pk}$. For icosahedron, the threshold $th \approx 1.29107$ [1 - 3] is used in the 3D local descriptor-based abnormality detection in traffic surveillance video [1]. Then th is subtracted from the projected vector $\widehat{\mathbf{pB}_\mathbf{B}}$ and negative values are set to 0.

Finally, the gradient magnitude value is binned as follows:

$$PB = \frac{\|G_B\|_2 \cdot \widehat{p_B}}{\|\widehat{p}_B\|_2} \quad \textbf{(8)}$$

where $\widehat{\mathbf{p}'_\mathbf{B}}$ is a thresholded histogram.

• **Histogram computation**: A given video cube is divided into cells, and each cell is further subdivided into blocks. Let a cell be represented as

$$V_C = \left[x_{V_C}, y_{V_C}, t_{V_C}, w_{V_C}, h_{V_C}, l_{V_C}\right]^T \quad \textbf{(9)}$$

where *x, y,* and *t* represent the position of Cth cell of video and *w, h, l* are width, height, and temporal length of Cth cell. This cell is divided into R*R * R sub-blocks B_j. For each sub-block B_j, there is a mean gradient G¯_(B_j) is obtained by

Eq.(12). Then P_{Bj} is quantized and represented as $\mathbf{P_{Bj}}$. The histogram H_{vc} for cell V_c is obtained by submission of quantized mean gradient vectors of all sub-blocks of that particular cell, and can be given as follows:

$$H_C = \sum_{j=1}^{R^3} \mathbf{p_{B_j}}$$

Then, finally, by concatenating these histograms, a 3D-HOG feature vector is obtained for a particular video cube, and the 3D-HOG feature vector is represented as follows:

$$HOG_A = \left(H_{C_1}, H_{C_2}, \ldots, H_{C_N}\right) \quad \textbf{(10)}$$

3D Histogram of Optical Flow (3D-HOOF)

For the computation of 3D-HOOF, optical flow for each frame of the video is obtained using the Lucas-Kanade algorithm. Then, optical flow is represented as an image, and the same process is followed as for obtaining the 3D-HOG, and we obtain a 3D-HOOF feature vector.

$$HOOF_A = (HOOF_{C_1}, HOOF_{C_2} \ldots., \ \mathrm{HOOF}_{C_N}\ ) \quad \textbf{(11)}$$

The final representation of the video cube can be obtained by concatenation of both feature vectors and can be represented as [$HOG_A HOOF_A$], which consists of appearance information extracted by 3D-HOG and motion information extracted by 3D-HOOF.

Classification

After extracting features from the video, pattern-finding is performed to analyze the features of each frame and identify patterns that can distinguish between anomalous and normal events or expected behaviour. Real-world scenes have much more complex distributions of features; a single classifier may not be able to model them correctly, but combining multiple classifiers can model such complex scenes effectively. Because each classifier focuses on a selected set of features or data distribution, combining two or more classifiers leverages the strengths of each [5 - 7]. Based on this evidence, our proposed algorithm uses a combined classifier to separate abnormal and normal events in surveillance video. Our proposed algorithm uses SVDD and a K-Means classifier. According to a previous studies [8], SVDD can more accurately identify outliers even though it has not used any negative examples during training. In comparison, K-means has the advantage of scale for large datasets, along with better generalization to clusters of different sizes and shapes [24 - 26].

SVDD (Support Vector Data Descriptor): A good data descriptor encompasses all the target data points while avoiding any superfluous space. The boundary in the dataset is used to find outliers present in the dataset. SVDD [9 - 11] creates a spherical boundary around the dataset, similar to the separating plane used in support vector machines. SVDD is more flexible due to the use of a kernel function. A simple representation of the SVDD classifier is shown in Fig. (**5**).

Let us consider a training *set* x_i *=[1,2, ..., N]* for which we need to create a boundary, where vector x is a column vector and $x^2 = x.x.$. SVDD creates a closed boundary (a hypersphere) around the training data. This hypersphere can be described by its center c and having radius R greater than 0. SVDD tries to minimize the volume of the hypersphere by minimizing the radius of the hypersphere (R^2) considering that all the training data points are well within the sphere. According to previous studies [9 - 12], the error function, which is to be minimized, can be given as:

$$\min_{c,R,\zeta} R^2 + C \sum_i \zeta_i \quad \textbf{(12)}$$

This is subjected to:

$$\|x_i - c\|^2 \leq R^2 + \zeta_i, \quad \textbf{(13)}$$

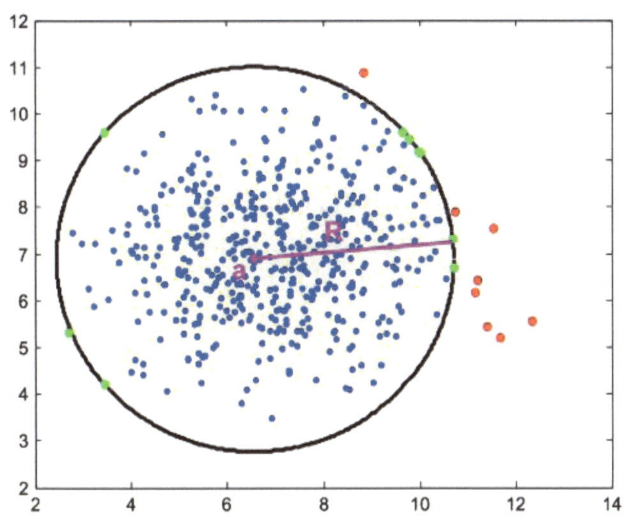

Fig. (5). Support vector data description [5] of 3D local descriptor-based abnormality detection in traffic surveillance video.

$$i \in [1,2 \ldots N]\zeta_i \geq 0$$

Where *C* is the tradeoff parameter that controls the balance between the errors and the volume of the sphere (R^2). ζ is a slack variable.

This minimization function is solved using Lagrangian multipliers.

$$L(R, c, \alpha_i, \gamma_i, \zeta_i) = R^2 + C\sum_i \zeta_i - \sum_i \alpha_i\{R^2 + \zeta_i - (\|x_i\|^2 - 2c.x_i + \| c \|^2)\} - \sum_i \gamma_i\zeta_i \quad \textbf{(14)}$$

where $\alpha_i \geq$ and $\gamma_i \geq 0$ are the Lagrange multipliers. L should be maximized with respect to R, c, ζ_i and minimized with respect to α_i and γ_i. The original optimization problem becomes the dual problem of Lagrange, and can be given as:

$$\min_{\alpha\geq 0,\gamma\geq 0}\left(\max_{a,R,\zeta} L(a, R, \zeta, \alpha, \gamma)\right) \quad \textbf{(15)}$$

Setting partial derivatives to 0 gives the constraints:

$$\frac{\partial L}{\partial R} = 0 \Rightarrow R(1 - \textstyle\sum_{i=1}^{N} \alpha_i) = 0 \Rightarrow \sum_{i=1}^{N} \alpha_i = 1$$
$$\frac{\partial L}{\partial a} = 0 \Rightarrow a = \frac{\sum_{i=1}^{N} \alpha_i}{\sum_i^N \alpha_i} = 0 \Rightarrow \textstyle\sum_{i=1}^{N} \alpha_i x_i \quad \textbf{(16)}$$
$$\frac{\partial L}{\partial \xi_i} = 0 \Rightarrow C - \alpha_i - \gamma_i = 0$$

from Eq.(16), $\alpha_i = C - \gamma_i$ and $\alpha_i \geq$ and $\gamma_i \geq 0$ γ_i can be removed, and these become:

$$0 \leq \alpha_i \geq C \quad \textbf{(17)}$$

After substituting Eq.(16)-Eq.(17) into Eq.(14)

$$L = \textstyle\sum_i^N \alpha_i(x_i \cdot x_i) - \sum_{i,j}^N \alpha_i\alpha_j(x_i . x_j) \quad \textbf{(18)}$$

Subjected to Eq.(17), maximizing Eq.(18) gives a set α_i. The Lagrange multiplier $\alpha_i = 0$, when an object x_i satisfies the inequality $\| x_i - c\|^2 < R^2 + \zeta_i$. When an object x_i satisfies the equality $\| x_i - c\|^2 < R^2 + \zeta_i$, α_i will become greater than 0 ($\alpha_i > 0$).

$$\|x_i - c\|^2 < R^2 \Rightarrow \alpha_i = 0, \gamma_i = 0$$
$$\|x_i - c\|^2 = R^2 \Rightarrow 0 < \alpha_i < C, \gamma_i = 0 \quad \textbf{(19)}$$
$$\|x_i - c\|^2 > R^2 \Rightarrow \alpha_i = C, \gamma_i > 0$$

A test object is only accepted when its distance from the test object and the sphere's center is smaller than or equal to the radius.

$$\| t - c \|^2 = (t.t) - 2\sum_i^N \alpha_i (t.x_i) + \sum_{i,j} \alpha_i \alpha_j (x_i.x_j) \le R^2 \quad \textbf{(20)}$$

R^2 is the distance between any support vector and the center of the hypersphere. Inner products of objects (x_i. x_j) can be replaced by a Gaussian kernel function K_G (x_i. x_j) $\exp^- = (x_i.\ x_j)^2/\sigma^2$.

K-Means classifier: K-means data description represents the data using K clusters, positioned such that the average distance to each cluster center is minimized. The cluster centers c_i are placed using the standard k-means clustering procedure [13]. The target class is then characterized by:

$$f(x) = \min_i (x - c_i)^2 \quad \textbf{(21)}$$

Then the output of the classifier can be given as:

$$O(x) = \begin{cases} \text{Normal} & f(x) \le \theta \\ \text{Abnormal} & f(x) > \theta \end{cases} \quad \textbf{(22)}$$

To make the outputs of the different classifiers comparable, confidence or posterior probabilities are estimated. Posterior probabilities (P_m (t)) for each test object *t* for every computed for every one of the *N* classes on which the 3D local descriptor-based abnormality detection in traffic surveillance video classifiers are trained, and is constrained to lie between 0 and 1. The posterior probabilities are normalized such that:

$$\sum_m^N P_m(t) = 1 \quad \textbf{(23)}$$

The values of the number of clusters in -means are 20, and the value of σ is 3 for the Gaussian kernel width in SVDD after experimenting with multiple values. For these values of K and σ, this algorithm gets satisfactory results on YouTube video [26 - 30].

RESULTS AND DISCUSSION

Our proposed algorithm is evaluated on CCTV surveillance videos from YouTube, as we could not find a standard dataset for detecting anomalies in traffic surveillance videos. This surveillance video was shot by a CCTV camera facing the road. This video has some frames of normal traffic, as well as some frames showing abnormality, *e.g.*, an accident on the road, a traffic jam, and an illegal U-turn. The frame size of the video is 352×240. Some frames of the video taken from YouTube are shown in Fig. (**6**).

Fig. (6). Some frames of the video used for the validation of our proposed Algorithm.

Where the upper row of the figure shows normal traffic scenes, while the lower row shows abnormal traffic scenes. The performance of our proposed algorithm is evaluated through the ROC curve and Accuracy for three different sets of classifiers. In the first experiment, only SVDD is used as a classifier, whereas for the second experiment, only K-Means is used. In the third experiment, an ensemble of SVDD and K-means is used for the purpose of classification. AUC and accuracy values for the three experimental settings are listed in Table **1**. The values reported here are comparable to similar experiments in the literature and, in this work, for other anomaly detection scenarios, confirming the performance of the proposed algorithm for anomaly detection in traffic surveillance videos. As mentioned earlier, due to the unavailability of a standard dataset for traffic surveillance, a direct comparison of the proposed algorithm with other algorithms could not be conducted. Another noteworthy outcome of these experiments is that the combined classifier outperformed the individual classifiers, *i.e.*, SVDD and K-means. Fig. (7) shows the detection results obtained using the combined classifier.

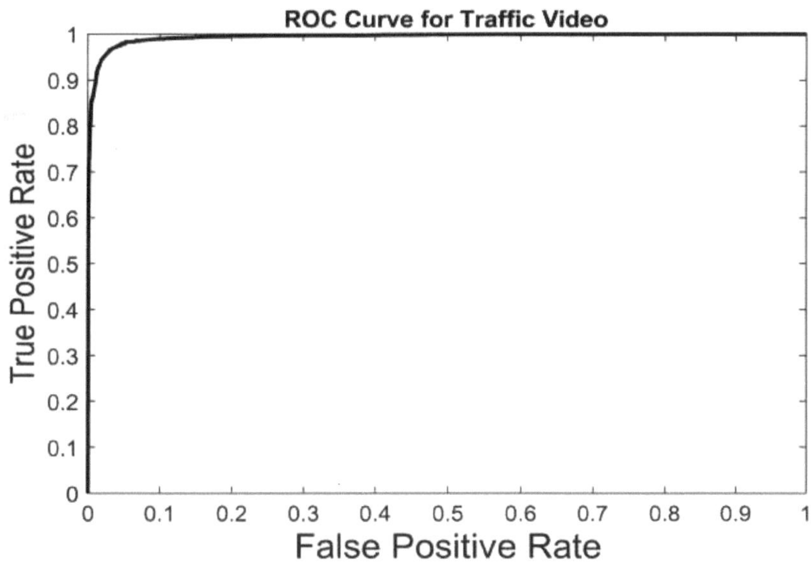

Fig. (7). ROC Curve of Combined Classifier.

Table 1. AUC and Accuracy of different classifiers.

S.N.	Method	AUC	Accuracy
1	SVDD Classifier	0.9906	95.13
2	K-Means Classifier	0.9890	93.75
3	Combined Classifier (SVDD-K-Means)	0.9941	96.33

CONCLUSION

In this work, an effective framework for detecting abnormality in traffic surveillance video is proposed, leveraging appearance and motion features. Both features describe the scene effectively. The proposed method exploits the advantage of combining multiple classifiers to achieve good performance, *viz.* SVDD and K-means. The 3D local descriptor-based abnormality detection results on the dataset show the performance of our proposed algorithm. Our proposed algorithm achieves a good AUC value as well as good accuracy for the video taken from YouTube.

REFERENCE

[1] A. Klaser, M. Marszalek, and C. Schmid, *A spatio-temporal descriptor based on 3d-gradients.*, 2008.

[2] R. Chalapathy, A.K. Menon, and S. Chawla, "201a). Robust, deep and inductive anomaly detection", *Joint European Conference on Machine Learning and Knowledge Det. co very in Database,* pp. 36-51

[3] Qiulei Dong, Yihong Wu, and Zhanyi Hu, "Pointwise motion image (pmi): A novel motion representation and its applications to abnormality detection and behavior recognition", *IEEE Trans. Circ. Syst. Video Tech.,* vol. 19, no. 3, pp. 407-416, 2009. [http://dx.doi.org/10.1109/TCSVT.2009.2013503]

[4] K.K. Santhosh, D.P. Dogra, and P.P. Roy, "Anomaly detection in road traffic using visual surveillance: A survey", *ACM Comput. Surv.,* vol. 53, no. 6, pp. 1-26, 2021. [http://dx.doi.org/10.1145/3417989]

[5] R. Kumar, R.K. Ratnesh, R.K. Chauhan, A. Kumar, M.K. Singla, and R. Gupta, "Non-invasive bio-impedance imaging and sensing for medical diagnostics and industrial applications", *J. Electrochem. Soc.,* vol. 171, no. 10, p. 107510, 2024. [http://dx.doi.org/10.1149/1945-7111/ad830b]

[6] D.M. Tax, and R.P. Duin, "Combining one-class classifiers", *International Workshop on Multiple Classifier Systems,* pp. 299-308, 2001. [http://dx.doi.org/10.1007/3-540-48219-9_30]

[7] M. Asad, J. Yang, E. Tu, L. Chen, and X. He, "Anomaly3D: Video anomaly detection based on 3D-normality clusters", *J. Vis. Commun. Image Represent.,* vol. 75, p. 103047, 2021. [http://dx.doi.org/10.1016/j.jvcir.2021.103047]

[8] R. Kumar, and S. Tripathi, "A Novel GUI-Based Image Reconstruction Algorithm of EIT Imaging Technique", *Int. J. Cogn. Inform. Nat. Intell.,* vol. 15, no. 3, pp. 31-46, 2021. [IJCINI]. [http://dx.doi.org/10.4018/IJCINI.20210701.oa3]

[9] T. Jan, M. Piccardi, and H. Gunes, "Suspicious behavior assessment for visual surveillance using neural network classifiers", In: *CISST.*, 2003, pp. 657-664.

[10] J.A.K. Suykens, and J. Vandewalle, "Least squares support vector machine classifiers", *Neural Process. Lett.,* vol. 9, no. 3, pp. 293-300, 1999. [http://dx.doi.org/10.1023/A:1018628609742]

[11] R. Maqsood, U.I. Bajwa, G. Saleem, R.H. Raza, and M.W. Anwar, "Anomaly recognition from surveillance videos using 3D convolution neural network", *Multimedia Tools Appl.,* vol. 80, no. 12, pp. 18693-18716, 2021. [http://dx.doi.org/10.1007/s11042-021-10570-3]

[12] S. Zheng, "Smoothly approximated support vector domain description", *Pattern Recognit.,* vol. 49, pp. 55-64, 2016. [http://dx.doi.org/10.1016/j.patcog.2015.07.003]

[13] E. Akdag, G. D'Amicantonio, J. Vijverberg, D. Stajan, B. Beers, P.H.N. De With, and E. Bondarev, "Geo-spatial traffic behaviour analysis and anomaly detection for ITS applications", *IET Intell. Transp. Syst.,* vol. 18, no. S1, pp. 2939-2962, 2024. [http://dx.doi.org/10.1049/itr2.12591]

[14] D.M.J. Tax, and R.P.W. Duin, "Support vector data description", *Mach. Learn.,* vol. 54, no. 1, pp. 45-66, 2004. [http://dx.doi.org/10.1023/B:MACH.0000008084.60811.49]

[15] A. Bin Queyam, M. Aljaidi, N. K. Bhoi, R. Kumar, & M. K. Singla, (2025, June). "Machine learning pipeline for bearing fault diagnosis using statistical and spectral features". In IET Conference Proceedings CP933 (Vol. 2025, No. 18, pp. 103-110). Stevenage, UK: The Institution of Engineering and Technology. (IET-2025).

[16] T. Reitmaier, and B. Sick, "The responsibility weighted mahalanobis kernel for semi- supervised training of support vector machines for classification", *Sciences (New York),* vol. 323, pp. 179-198, 2015.

[17] M.I. Sarker, C. Losada-Gutiérrez, M. Marrón-Romera, D. Fuentes-Jiménez, and S. Luengo-Sánchez, "Semi-supervised anomaly detection in video-surveillance scenes in the wild", *Sensors (Basel),* vol. 21, no. 12, p. 3993, 2021.
[http://dx.doi.org/10.3390/s21123993] [PMID: 34207883]

[18] J. A. Suykens, and J. Vandewalle, "Least squares support vector machine classifiers", *Neural proce sin,g letter,* vol. 9, no. 3, p. 293p. 300, 1999.

[19] D. Raghav, D. Shekhawat, A. Bin Queyam, R. Kumar, & M. K. Singla, (2025, June). "Enhancing communication for disabled individuals with an eye-blink operated virtual keyboard". In IET Conference Proceedings CP933 (Vol. 2025, No. 18, pp. 2-7). Stevenage, UK: The Institution of Engineering and Technology. (IET-2025).

[20] R. Kumar, R. K. Ratnesh, P. Gupta, M. K. Singla, and P. K. Pandey, “IoT-integrated bio-impedance monitoring for enhanced agricultural applications,” *J. Electrochem. Soc.*, vol. 172, no. 4, 2025.

[21] K. Chidananda, and A.P.S. Kumar, "A Robust Multi Descriptor Fusion with One-Class CNN for Detecting Anomalies in Video Surveillance", *Int. J. Saf. Secur. Eng.,* vol. 13, no. 6, pp. 1143-1151, 2023.
[http://dx.doi.org/10.18280/ijsse.130618]

[22] G.M. Bishop, *Neural networks for pattern recognition* Oxford university press, 1995.

[23] M. Asad, H. Jiang, J. Yang, E. Tu, and A.A. Malik, "Multi-Stream 3D latent feature clustering for abnormality detection in videos", *Appl. Intell.,* vol. 52, no. 1, pp. 1126-1143, 2022.
[http://dx.doi.org/10.1007/s10489-021-02356-9]

[24] X. Hu, J. Lian, D. Zhang, X. Gao, L. Jiang, and W. Chen, "Video anomaly detection based on 3D convolutional auto-encoder", *Signal Image Video Process.,* vol. 16, no. 7, pp. 1885-1893, 2022.
[http://dx.doi.org/10.1007/s11760-022-02148-9]

[25] K.U. Duja, I.A. Khan, and M. Alsuhaibani, "Video Surveillance Anomaly Detection: A Review on Deep Learning Benchmarks", *IEEE Access,* vol. 12, pp. 164811-164842, 2024.
[http://dx.doi.org/10.1109/ACCESS.2024.3491868]

[26] S. Anoopa, and A. Salim, "Survey on anomaly detection in surveillance videos", *Mater. Today Proc.,* vol. 58, pp. 162-167, 2022.
[http://dx.doi.org/10.1016/j.matpr.2022.01.171]

[27] Y. Yao, M. Xu, Y. Wang, D.J. Crandall, and E.M. Atkins, "Unsupervised traffic accident detection in first-person videos", *2019 IEEE/RSJ International Conference on Intelligent Robots and Systems (IROS),* 2019pp. 273-280
[http://dx.doi.org/10.1109/IROS40897.2019.8967556]

[28] R.T. Merlin, R. Karthick, A.A. Babu, G.V. Selvi, D. Usha, and R. Nithya, "Abnormal events detection using spatio-temporal saliency descriptor and fuzzy representation analysis", *Sci. Rep.,* vol. 14, no. 1, p. 29818, 2024.
[http://dx.doi.org/10.1038/s41598-024-81387-x] [PMID: 39616248]

[29] R. Raja, P.C. Sharma, M.R. Mahmood, and D.K. Saini, "Analysis of anomaly detection in surveillance video: recent trends and future vision", *Multimedia Tools Appl.,* vol. 82, no. 8, pp. 12635-12651, 2023.
[http://dx.doi.org/10.1007/s11042-022-13954-1]

[30] R. Kumar, R. K. Ratnesh, J. Singh, A. Kumar, and R. Chandra, “IoT-driven experimental framework for advancing electrical impedance tomography,” *ECS J. Solid State Sci. Technol.*, vol. 13, no. 2, p. 027002, 2024.

CHAPTER 5

Estimating the Energy of Low-Quality Images Using Kinetic Energy and a Hybrid Model

M. Bhanurangarao[1], **D. V. Naga Raju**[1], **Meduri Raghuchandra**[1], **Y. Yesu Jyothi**[1] and **M. Srikanth**[2,*]

[1] *Department of Information Technology, Shri Vishnu Engineering College for Women, Bhimavaram, AP, India*

[2] *Department of Information Technology, SRKR Engineering College, Bhimavaram, AP, India*

Abstract: Many applications rely on the correct analysis of low-quality photographs, yet effective interpretation is sometimes hampered by the inherent constraints of degraded image settings. We present a novel approach to estimating the energy content of low-quality images. We create a framework for measuring the dynamic features of these images by applying physical rules, particularly those governing motion. To preprocess the photos for precise energy estimation, we use image enhancement techniques and motion features. Following that, we calculate the energy of the objects' motions in the images using kinetic energy principles, considering characteristics, such as mass and velocity. Our proposed method accurately calculates energy levels from low-quality pictures, as demonstrated experimentally with synthetic and real-world datasets. This methodology has applications in environmental monitoring, robotics, and surveillance, all of which rely on insights into energy dynamics for analysis and decision-making. It also increases our understanding of the dynamics involved in such imaging.

Keywords: Energy estimation, Hybrid model, Image enhancement, Kinetic energy, Low-quality images, Motion features.

INTRODUCTION

Image analysis relies heavily on high-quality input data to produce accurate and reliable results. However, there are many practical applications that rely on interpreting low-resolution images, which pose significant challenges due to their inherent limits. Extracting valuable information from these images, often degraded by factors like poor lighting, blurred movement, or poor resolution,

* **Corresponding author M. Srikanth:** Department of Information Technology, SRKR Engineering College, Bhimavaram, AP, India; E-mail: srikanth.mandela@gmail.com

requires ingenuity. Crucial to this study is the estimation of the energy content of these images, which may shed light on the depicted scenes' dynamics [1].

Traditional methods for calculating image energy rely heavily on high-quality visual data and can fall short when confronted with intricate picture degradation configurations. We propose a novel method to fill this void by applying the principles of kinetic energy to estimate an object's energy content in low-resolution images. To ensure precise energy estimations, our system preprocesses and refines the visual input using a combination of motion feature analysis and image enhancement techniques [2 - 4].

Our methodology is based on the principles of kinetic energy laws, which govern the general laws of motion. By examining the mass and velocity of the objects in the images, we can precisely ascertain their energy levels. This hybrid model does double duty: it increases the interpretability of low-quality photos and broadens their typical uses [5 - 8].

To rigorously test our proposed method on both synthetic and real-world datasets, its efficacy in various contexts is demonstrated. Possible uses of robotics, surveillance, and environmental monitoring for accurately measuring energy from low-quality images are substantial. Because knowledge of energy dynamics is critical to analysis and decision-making in various domains, our method represents a major advancement in the field. The sections that follow provide a detailed explanation of our framework, including its motion feature analysis, image enhancement techniques, and kinetic energy principles. Also, the effectiveness of our technology is demonstrated through experiments, and its potential uses are discussed. We aim to enhance energy estimation in low-quality images, enabling more dependable and robust analysis across diverse technical and scientific fields.

IMAGE ENHANCEMENT TECHNIQUES

Features from low-quality photos are difficult to extract because of common issues like noise, poor contrast, and blurring. Our approach to resolving these issues is based on a set of picture enhancement methods targeted at raising the overall quality of the images. To minimise noise while preserving important information, we use state-of-the-art denoising methods such as non-local means (NLM) and Gaussian filtering. We improve object-to-background separation by increasing contrast through adaptive histogram equalisation. To enhance the photographs, we also use sharpening filters, such as unsharp masking, to emphasize the edges and finer details. By changing parameters such as the kernel size and the sigma of the adaptive histogram equalisation, we can achieve a good balance between removing noise and preserving features. Reduced noise, improved contrast, and better fine-tuning yield a more suitable set of photos for processing and motion feature extraction [4 - 8].

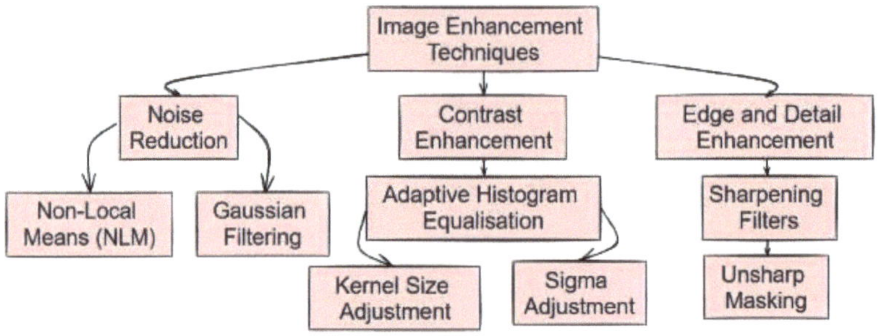

Fig. (1). Image enhancement techniques for low-quality photographs, illustrating various methods such as noise reduction, contrast adjustment, and edge sharpening to improve image clarity before kinetic energy estimation.

Low-quality photographs often exhibit noise, poor contrast, and blurring; these shortcomings make it difficult to extract features from the images. Our approach to solving these issues begins with a series of image enhancement techniques designed to improve the overall image quality. We employ the most advanced denoising techniques, such as non-local means (NLM) and Gaussian filtering, to reduce background noise while keeping crucial information, as shown in Fig.**1** [9 - 13]. Using adaptive histogram equalisation, we can improve object-to-background separation by increasing contrast. To bring out the best in the photos by emphasizing edges and minute details, we also apply sharpening filters, such as unsharp masking. By adjusting parameters, such as the sigma and kernel size of the adaptive histogram equalisation technique, we can effectively balance noise reduction and feature preservation. The final collection of images improves with reduced noise, increased contrast, and finer details, making it suitable for motion feature extraction and processing [14 - 16].

The dataset contains a wide range of low-quality images with varying levels of brightness, contrast, and noise, which makes feature extraction challenging. We assign a unique Image ID to each image based on its initial values for sharpness (Very Blurry, Blurry, Slightly Blurry), contrast (Very Low, Low, Medium, High, Very High), and noise level (High, Medium, Low), as shown in Table **1**. To enhance these images, we employ advanced denoising techniques, such as non-local means (NLM) and Gaussian filtering, as well as fine-tuning parameters, such as sigma and kernel size, to strike a fair balance between minimizing noise and maintaining clarity. We use adaptive histogram equalisation to enhance contrast and improve object-to-background separation by adjusting the clip limit and grid size parameters. We highlight edges and small details by using sharpening filters with specific settings for quantity, radius, and threshold, such as unsharp masking. The significant gains in noise level, contrast, and sharpness in the dataset enable better motion feature extraction and processing.

Table 1. Image Enhancement Techniques for Low-Quality Photographs, outlining various methods such as noise reduction, contrast adjustment, and edge sharpening, along with their application and effectiveness in improving image quality before kinetic energy estimation.

Image ID	Noise Level	Initial Contrast	Initial Sharpness	Denoising Method	Denoising Parameters	Contrast Enhancement Method	Contrast Enhancement Parameters	Sharpening Method	Sharpening Parameters	Final Noise Level	Final Contrast	Final Sharpness
Img001	High	Low	Blurry	NLM	(7, 0.1)	Adaptive Histogram Equalization	Clip Limit: 2.0, Grid Size: 8x8	Unsharp Masking	(1.5, 1.0, 0)	Low	High	Sharp
Img002	Medium	Medium	Slightly Blurry	Gaussian	(5, 0.2)	Adaptive Histogram Equalization	Clip Limit: 1.5, Grid Size: 8x8	Unsharp Masking	(1.2, 0.5, 0)	Low	High	Sharper
Img003	Low	High	Very Blurry	NLM	(9, 0.3)	Adaptive Histogram Equalization	Clip Limit: 2.5, Grid Size: 8x8	Unsharp Masking	(2.0, 1.5, 0)	Very Low	Very High	Very Sharp
Img004	High	Very Low	Blurry	Gaussian	(7, 0.4)	Adaptive Histogram Equalization	Clip Limit: 3.0, Grid Size: 8x8	Unsharp Masking	(1.7, 1.0, 0)	Medium	Medium	Sharp
Img005	Medium	Low	Slightly Blurry	NLM	(5, 0.2)	Adaptive Histogram Equalization	Clip Limit: 1.8, Grid Size: 8x8	Unsharp Masking	(1.3, 0.7, 0)	Low	High	Sharper
Img006	Low	Medium	Blurry	Gaussian	(3, 0.1)	Adaptive Histogram Equalization	Clip Limit: 1.2, Grid Size: 8x8	Unsharp Masking	(1.1, 0.4, 0)	Very Low	High	Sharp
Img007	High	Low	Very Blurry	NLM	(8, 0.5)	Adaptive Histogram Equalization	Clip Limit: 2.2, Grid Size: 8x8	Unsharp Masking	(1.8, 1.2, 0)	Medium	High	Very Sharp
Img008	Medium	Very Low	Slightly Blurry	Gaussian	(6, 0.3)	Adaptive Histogram Equalization	Clip Limit: 2.0, Grid Size: 8x8	Unsharp Masking	(1.4, 0.8, 0)	Low	Medium	Sharper
Img009	Low	Medium	Blurry	NLM	(4, 0.2)	Adaptive Histogram Equalization	Clip Limit: 1.7, Grid Size: 8x8	Unsharp Masking	(1.2, 0.6, 0)	Very Low	High	Sharp
Img010	High	Low	Very Blurry	Gaussian	(8, 0.6)	Adaptive Histogram Equalization	Clip Limit: 2.8, Grid Size: 8x8	Unsharp Masking	(2.1, 1.3, 0)	Medium	High	Very Sharp

We address common issues such as noise, low contrast, and blurring in low-quality photos by combining a range of cutting-edge image enhancement techniques. These methods include denoising with Gaussian filtering, sharpening with unsharp masking, denoising with adaptive histogram equalisation, and non-local means (NLM). The following formula captures the effectiveness of these enhancements by combining the key components of these techniques.

$$I_{enhanced}=S(U(H(N(I,k_n,\sigma_n),k_h,\sigma_h,clip,grid)),k_s,a_s,r_s,t_s) \quad \textbf{(1)}$$

The sequential application of processing techniques, including denoising, contrast enhancement, and sharpening, results in an improved version of the original image. Non-Local Means (NLM), also known as Gaussian filtering with kernel size k_n and sigma σ_n, is used in denoising to get rid of noise while keeping important features. Adaptive histogram equalisation is applied to the denoised image in the contrast enhancement stage H(N,kh,σh,clip,grid) to make it easier to distinguish between objects and backgrounds. The following parameters are used to achieve this equalisation: grid size h, sigma h, clip limit h, kernel k_h, and grid size h. Unsharp masking, which improves the edges and finer details of the contrast-enhanced image with parameters like kernel k_s, a_s an radius r_s, and threshold t_s, is the last step in the sharpening (Hprocess $U\ (H, k_s, a_s, r_s, t_s)$, as shown in Fig. (**2**). The technique accurately balances noise reduction, contrast enhancement, and detail sharpening by fine-tuning these settings, resulting in a higher-quality final image suitable for further processing and motion feature extraction.

Fig. (2a). Image enhancement techniques applied to low-quality photographs.

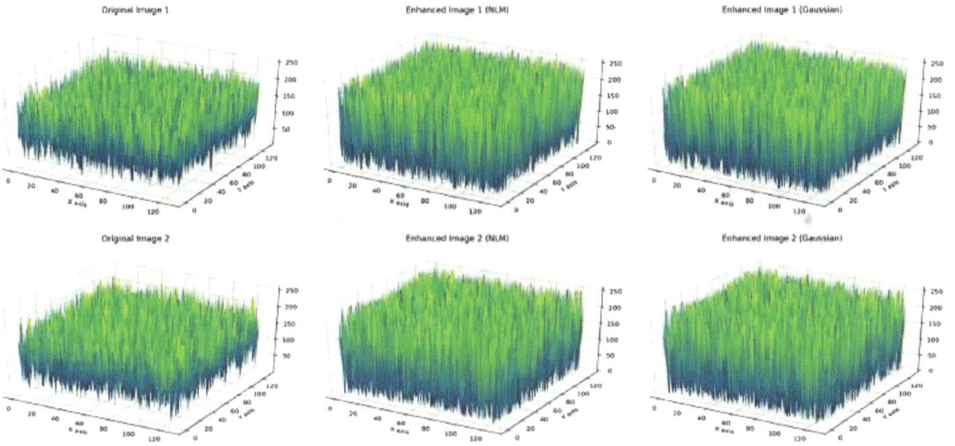

Fig. (2b). Image enhancement techniques applied to low-quality photographs, illustrating the processes of noise reduction, contrast adjustment, and edge sharpening to enhance image clarity before kinetic energy estimation.

MOTION FEATURE ANALYSIS

Identifying and tracking objects in low-quality images presents significant challenges due to issues, such as poor resolution, noise, and motion blur. To overcome these challenges, we utilise state-of-the-art object detection algorithms, such as YOLO (You Only Look Once) and Faster R-CNN (Region Convolutional Neural Network), to identify objects in each frame accurately. Following object detection, we apply optical flow techniques, specifically the Lucas-Kanade method, to estimate the velocity of the detected objects by analysing their positional changes over time. Key parameters for this process include confidence thresholds for object detection to minimise false positives and negatives, as well as configuration settings for optical flow calculations, such as window size and pyramid levels, to ensure accurate motion estimation. Effective object detection and tracking enable precise measurement of object velocities, which is crucial for kinetic energy estimation [17 - 20]. The reliable extraction of motion features from enhanced images significantly improves the overall accuracy of our energy calculations, as shown in Fig. (**3**).

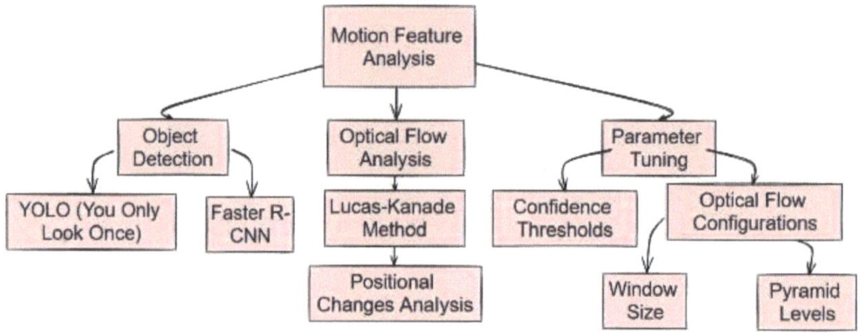

Fig. (3). Motion feature analysis in low-quality images, showing how motion features such as velocity, trajectory, and direction are extracted and analyzed to estimate kinetic energy, even in degraded image conditions.

This collection includes the parameters required for motion feature analysis in low-quality photos. Every row provides the following details to reduce the number of false positives and negatives: image ID, resolution, noise level, object detection method (YOLO or Faster R-CNN), and object detection confidence threshold. Additionally, you can view the number of items picked up in each frame. The dataset contains information on optical flow computations, namely the Lucas-Kanade method, with configurations, such as window size and pyramid levels. These details are required for a precise motion estimation, as shown in Table **2**. We also include the average pixel-per-frame velocity of the spotted objects. This large dataset is useful for testing object detection and tracking algorithms to estimate kinetic energy from low-quality images.

Table 2. Motion feature analysis in low-quality images, summarizing the extracted motion features such as velocity, trajectory, and direction of objects, along with their contribution to kinetic energy estimation.

Image ID	Reso lution	Noise Level	Object Detection Algorithm	Confid ence Thres hold	Detected Object Count	Optical Flow Method	Window Size	Pyramid Levels	Average Velocity (Pixels/Frame)
Img001	Low	High	YOLO	0.5	3	Lucas-Kanade	21	3	4.5
Img002	Medium	Medium	Faster R-CNN	0.7	5	Lucas-Kanade	15	4	3.2
Img003	Low	High	YOLO	0.6	4	Lucas-Kanade	19	3	5.1
Img004	High	Low	Faster R-CNN	0.8	6	Lucas-Kanade	17	2	2.8
Img005	Medium	Medium	YOLO	0.55	3	Lucas-Kanade	20	4	4
Img006	Low	High	Faster R-CNN	0.6	5	Lucas-Kanade	18	3	4.7

(Table 2) cont.....

Image ID	Reso lution	Noise Level	Object Detection Algorithm	Confid ence Thres hold	Detected Object Count	Optical Flow Method	Window Size	Pyramid Levels	Average Velocity (Pixels/Frame)
Img007	High	Low	YOLO	0.7	7	Lucas-Kanade	16	2	3.5
Img008	Medium	Medium	Faster R-CNN	0.75	4	Lucas-Kanade	14	4	2.9
Img009	Low	High	YOLO	0.65	3	Lucas-Kanade	21	3	5.3
Img010	Medium	Medium	Faster R-CNN	0.8	6	Lucas-Kanade	19	2	3

Detecting and following objects across a series of low-resolution images using optical flow analysis: At the outset, we generate simulated grayscale images and the coordinates of discovered objects, as shown in Fig. (**4**). Using the Lucas-Kanade method, the calculate_optical_flow function tracks the locations and velocities of observed objects between successive frames. The main function iteratively detects item mobility in the simulated images using optical flow computing. With every object it finds, it records its class ID, estimated velocity, and bounding box. This chapter defines but does not use the visualize_results function, which prints the results and graphically displays the bounding boxes and arrowsrepresentingt velocities on the photographs.

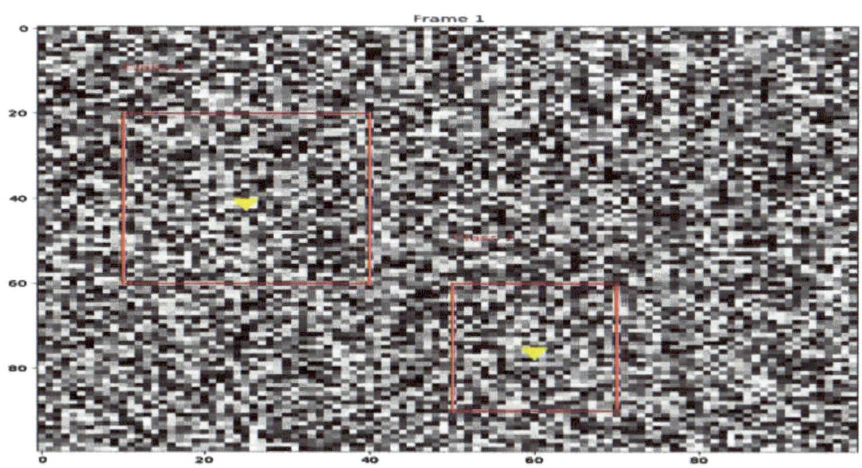

Fig. (4a). Detecting and following objects in a series of low-resolution images using optical flow analysis.

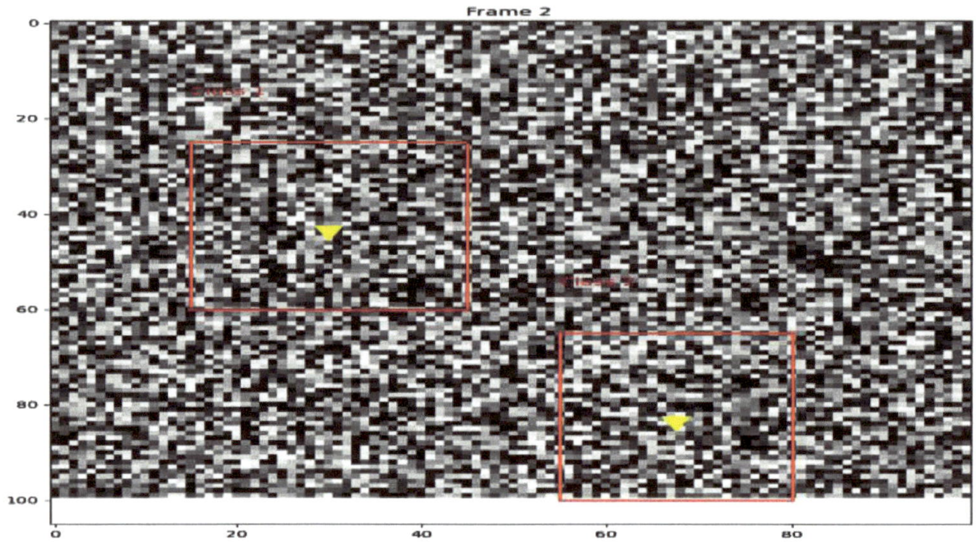

Fig. (4b). Detecting and following objects in a series of low-resolution images using optical flow analysis, illustrating the movement of objects across frames and the tracking of velocity and direction using optical flow techniques to estimate kinetic energy.

KINETIC ENERGY ESTIMATION

Estimating the kinetic energy of objects in low-quality images requires accurately determining their mass and velocity, which can be challenging given the inherent limitations of such images. To address this, we estimate the mass of detected objects using contextual information or predefined models relevant to the application domain, as shown in Fig. (**5**) [21 - 23]. We calculate the velocity of each object using data obtained from the optical flow analysis. The kinetic energy is then calculated using the fundamental physics formula $E_k=1/2mv^2$, where E_k is the kinetic energy, m is the mass, and v is the velocity. Parameters for this process include predefined mass values for common objects and velocity data derived from optical flow analysis [24 - 28]. By integrating these physical principles, we provide a reliable and meaningful estimation of the energy levels of objects in low-quality images, enhancing the interpretability and accuracy of our results.

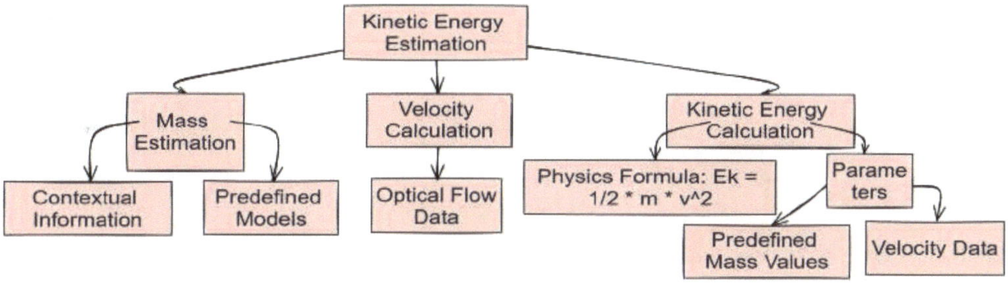

Fig. (5). Kinetic energy estimation from low-quality images, illustrating the process of calculating kinetic energy using motion features such as velocity and mass, extracted from degraded image data.

This dataset is used to estimate the kinetic energy of objects visible in low-resolution pictures. Each row in the table represents a distinct object in an image, with the aid of the object ID and the image ID. The Estimated Mass column gives the mass of each object as determined by either known models or contextual information, while the thing type column categorises the object (*e.g.*, car, pedestrian, bicycle, dog). An optical flow study results in the listing of the average velocity, expressed in meters per second (m/s). We can calculate the kinetic energy using the formula $f = 1/2mv^2$, where m and v represent the mass and velocity, respectively, as shown in Table **3**. This dataset facilitates kinetic energy estimation by improving the interpretability and accuracy of energy dynamics analysis in low-quality images, as shown in Fig. (**6**).

Table 3. Kinetic energy estimation from low-quality images, presenting the calculated kinetic energy for various objects based on their estimated mass and average velocity in low-resolution images.

Image ID	Object ID	Object Type	Estimated Mass (kg)	Average Velocity (m/s)	Kinetic Energy (J)
Img001	Obj001	Car	1200	4.5	12150
Img001	Obj002	Pedestrian	70	1.8	113.4
Img001	Obj003	Bicycle	15	3.2	76.8
Img002	Obj001	Dog	20	2.5	62.5
Img002	Obj002	Pedestrian	70	1.2	50.4
Img002	Obj003	Car	1500	3.8	10830
Img003	Obj001	Bicycle	15	4.1	125.9
Img003	Obj002	Pedestrian	70	2	140
Img004	Obj001	Car	1100	5.2	14872

(Table 3) cont.....

Image ID	Object ID	Object Type	Estimated Mass (kg)	Average Velocity (m/s)	Kinetic Energy (J)
Img004	Obj002	Dog	25	3	112.5
Img005	Obj001	Bicycle	15	3.5	91.9
Img005	Obj002	Car	1300	4	10400
Img006	Obj001	Pedestrian	70	1.5	78.8
Img006	Obj002	Bicycle	15	2.8	58.8
Img007	Obj001	Car	1400	3.9	10647
Img007	Obj002	Pedestrian	70	2.2	169.4
Img008	Obj001	Dog	20	2.7	72.9
Img008	Obj002	Bicycle	15	3.1	72.2
Img009	Obj001	Car	1600	4.4	15568
Img009	Obj002	Pedestrian	70	1.9	126.4
Img010	Obj001	Bicycle	15	3	67.5
Img010	Obj002	Dog	25	2.4	72

```
Img001   Obj001   Car         1200     4.50     12150.00
Img001   Obj002   Pedestrian           70       1.80         113.40
Img001   Obj003   Bicycle 15           3.20     76.80
Img002   Obj001   Dog           20     2.50     62.50
Img002   Obj002   Pedestrian           70       1.20         50.40
Img002   Obj003   Car         1200     3.80     8664.00
Img003   Obj001   Bicycle 15           4.10     126.07
Img003   Obj002   Pedestrian           70       2.00         140.00
Img004   Obj001   Car         1200     5.20     16224.00
Img004   Obj002   Dog           20     3.00     90.00
Img005   Obj001   Bicycle 15           3.50     91.88
Img005   Obj002   Car         1200     4.00     9600.00
Img006   Obj001   Pedestrian           70       1.50         78.75
Img006   Obj002   Bicycle 15           2.80     58.80
Img007   Obj001   Car         1200     3.90     9126.00
Img007   Obj002   Pedestrian           70       2.20         169.40
Img008   Obj001   Dog           20     2.70     72.90
Img008   Obj002   Bicycle 15           3.10     72.08
Img009   Obj001   Car         1200     4.40     11616.00
Img009   Obj002   Pedestrian           70       1.90         126.35
Img010   Obj001   Bicycle 15           3.00     67.50
Img010   Obj002   Dog           20     2.40     57.60
```

Fig. (6). Image details during processing, showing the sequential stages of image enhancement, motion feature extraction, and kinetic energy estimation applied to low-resolution images.

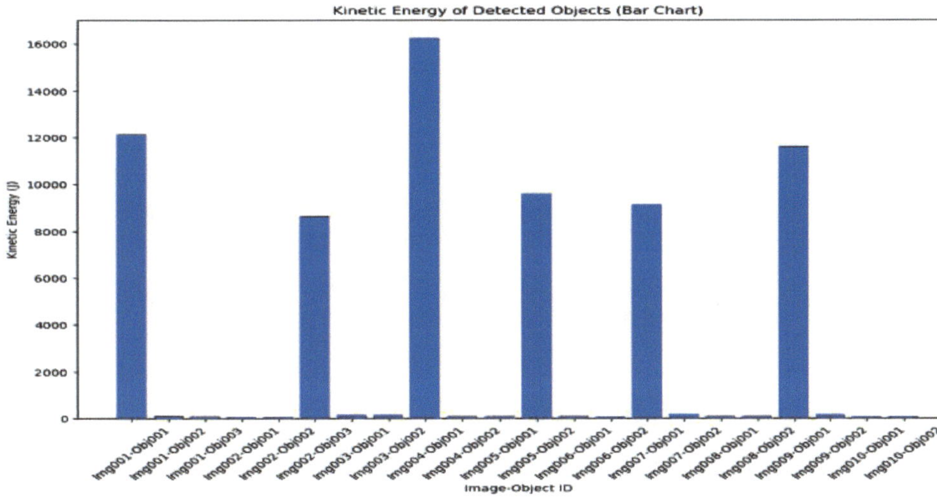

Fig. (7). Kinetic energy of detected objects, displaying the calculated kinetic energy for each object detected in the low-resolution images, based on their estimated mass and velocity.

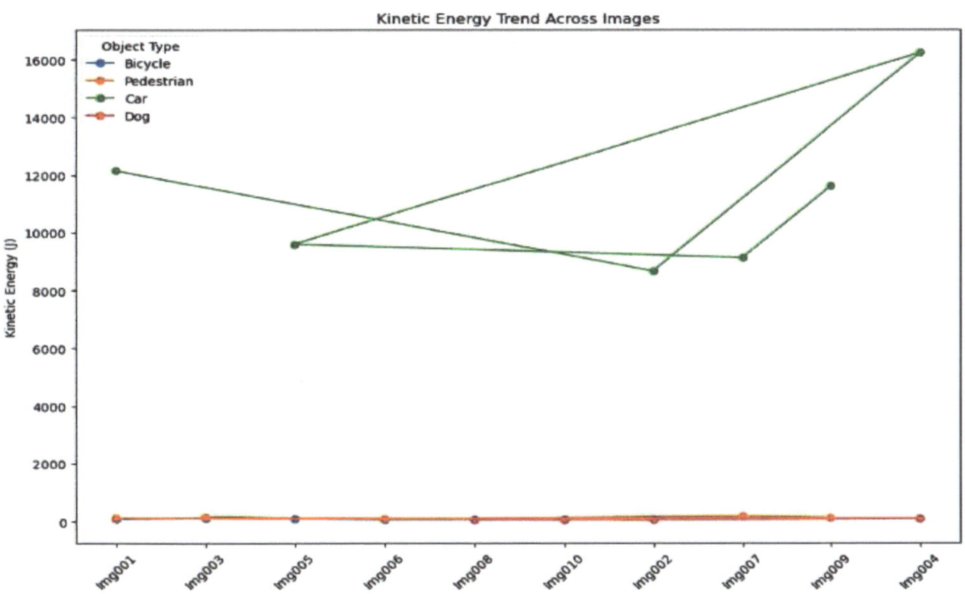

Fig. (8). Kinetic energy trend across images, illustrating the variation in kinetic energy for detected objects over a sequence of low-resolution images, showing how energy fluctuates based on object movement.

This describes a method for calculating and displaying the kinetic energy of objects in virtual images, incorporating the necessary libraries. Firstly, it establishes mass values for typical objects and models the detected items with their respective velocities. The function calculate_kinetic_energy uses the formula $f = 1/2mv^2$ to determine the kinetic energy. The visualize_kinetic_energy function generates a number of plots, including scatter plots, bar charts, line graphs, and histograms, to exhibit the kinetic energy data. In the main function, the resultsare obtained after looping through all identified objects and computing their kinetic energy using their velocities and predefined mass values. Finally, the visualization method yields the results as shown in Figs. (**7** and **8**). This method demonstrates the complete process of computing, inspecting, and displaying kinetic energy information for objects in a picture set.

HYBRID MODEL INTEGRATION

Purely physical models may not capture all the factors affecting energy estimation, especially given the variability in image quality and scene complexity. To address these limitations, we integrate a hybrid model that combines physical principles with machine learning techniques (Fig. **9**). This approach involves creating a comprehensive set of features that include physical attributes, such as mass and velocity, as well as additional contextual information extracted from the images. We then train a machine learning model using labelled datasets to learn the relationship between these features and the actual energy levels [28 - 31]. Parameters for this hybrid model include selecting suitable machine learning algorithms, such as regression models and neural networks, and defining a robust feature set for model training. Our hybrid model improves the accuracy and robustness of energy estimation by combining physical principles with machine learning. This lets it adapt to different situations and changes in image quality, making it more useful and general.

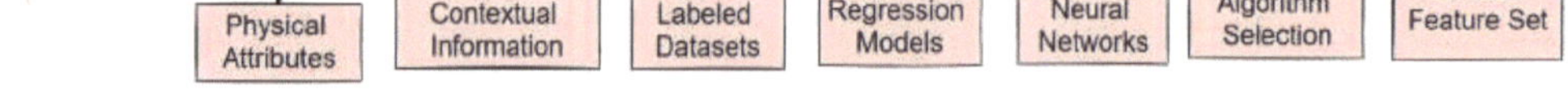

Fig. (9). Hybrid model integration for energy estimation, demonstrating the combination of image enhancement, motion feature analysis, and kinetic energy calculations within a unified framework for improved energy estimation from low-quality images.

Table 4. Hybrid model integration for energy estimation, summarizing the components and stages of the hybrid model, including image enhancement techniques, motion feature extraction, and kinetic energy calculation, and their roles in improving energy estimation accuracy.

Image ID	Object ID	Object Type	Estimated Mass (kg)	Average Velo city (m/s)	Initial Noise Level	Initial Contrast	Initial Sharpness	Final Noise Level	Final Contrast	Final Sharpness	Contextual Info (*e.g.*, Weather)	Actual Energy (J)
Img001	Obj001	Car	1200	4.5	High	Low	Blurry	Low	High	Sharp	Sunny	12150
Img001	Obj002	Pedestrian	70	1.8	High	Low	Blurry	Low	High	Sharp	Sunny	113.4
Img001	Obj003	Bicycle	15	3.2	High	Low	Blurry	Low	High	Sharp	Sunny	76.8
Img002	Obj001	Dog	20	2.5	Medium	Medium	Slightly Blurry	Low	High	Sharper	Cloudy	62.5
Img002	Obj002	Pedestrian	70	1.2	Medium	Medium	Slightly Blurry	Low	High	Sharper	Cloudy	50.4
Img002	Obj003	Car	1500	3.8	Medium	Medium	Slightly Blurry	Low	High	Sharper	Cloudy	10830
Img003	Obj001	Bicycle	15	4.1	Low	High	Very Blurry	Very Low	Very High	Very Sharp	Rainy	125.9
Img003	Obj002	Pedestrian	70	2	Low	High	Very Blurry	Very Low	Very High	Very Sharp	Rainy	140
Img004	Obj001	Car	1100	5.2	High	Very Low	Blurry	Medium	Medium	Sharp	Foggy	14872
Img004	Obj002	Dog	25	3	High	Very Low	Blurry	Medium	Medium	Sharp	Foggy	112.5
Img005	Obj001	Bicycle	15	3.5	Medium	Low	Slightly Blurry	Low	High	Sharper	Sunny	91.9
Img005	Obj002	Car	1300	4	Medium	Low	Slightly Blurry	Low	High	Sharper	Sunny	10400
Img006	Obj001	Pedestrian	70	1.5	Low	Medium	Blurry	Very Low	High	Sharp	Cloudy	78.8
Img006	Obj002	Bicycle	15	2.8	Low	Medium	Blurry	Very Low	High	Sharp	Cloudy	58.8
Img007	Obj001	Car	1400	3.9	High	Low	Very Blurry	Medium	High	Very Sharp	Rainy	10647
Img007	Obj002	Pedestrian	70	2.2	High	Low	Very Blurry	Medium	High	Very Sharp	Rainy	169.4
Img008	Obj001	Dog	20	2.7	Medium	Very Low	Slightly Blurry	Low	Medium	Sharper	Foggy	72.9
Img008	Obj002	Bicycle	15	3.1	Medium	Very Low	Slightly Blurry	Low	Medium	Sharper	Foggy	72.2
Img009	Obj001	Car	1600	4.4	Low	Medium	Blurry	Very Low	High	Sharp	Cloudy	15568
Img009	Obj002	Pedestrian	70	1.9	Low	Medium	Blurry	Very Low	High	Sharp	Cloudy	126.4
Img010	Obj001	Bicycle	15	3	High	Low	Very Blurry	Medium	High	Very Sharp	Sunny	67.5
Img010	Obj002	Dog	25	2.4	High	Low	Very Blurry	Medium	High	Very Sharp	Sunny	72

By combining physical attributes, contextual data, and machine learning techniques, this dataset aims to train and validate a hybrid model for energy estimation. Each entry represents an object present in a low-resolution image, complete with its unique Image ID and Object ID. Motion feature analysis derives the estimated mass and average velocity, among other physical attributes, and includes them in the dataset. We also provide measurements of image quality before and after the augmentation (noise level, contrast, and sharpness), as shown in Table**4**. We also add contextual data, such as the current weather, to give even more detail. We store the actual kinetic energy values in the real energy column to train the machine learning model (Fig. (**10**). By incorporating these diverse characteristics, the hybrid model improves energy estimation accuracy and robustness while accommodating variations in picture quality and scene complexity.

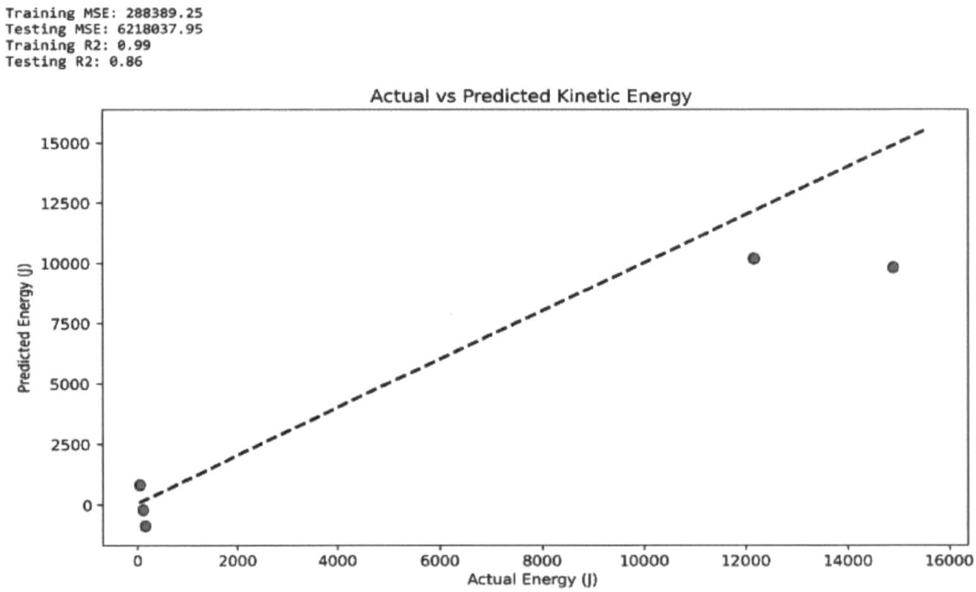

Fig. (10). Actual *vs* predicted kinetic energy, showing the comparison between the calculated kinetic energy and the expected values for both synthetic and real-world datasets.

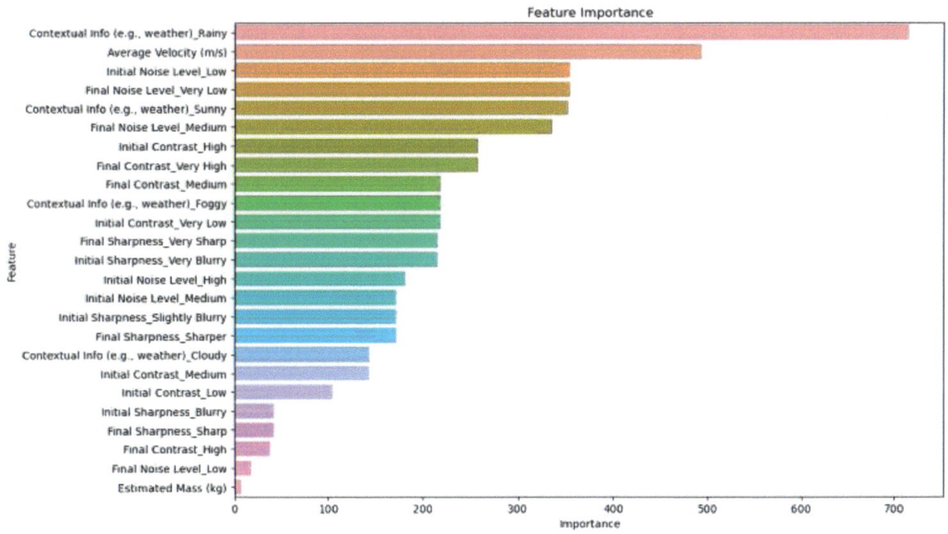

Fig. (11). Features of all objects in the image, including their properties, such as mass, velocity, and energy calculations.

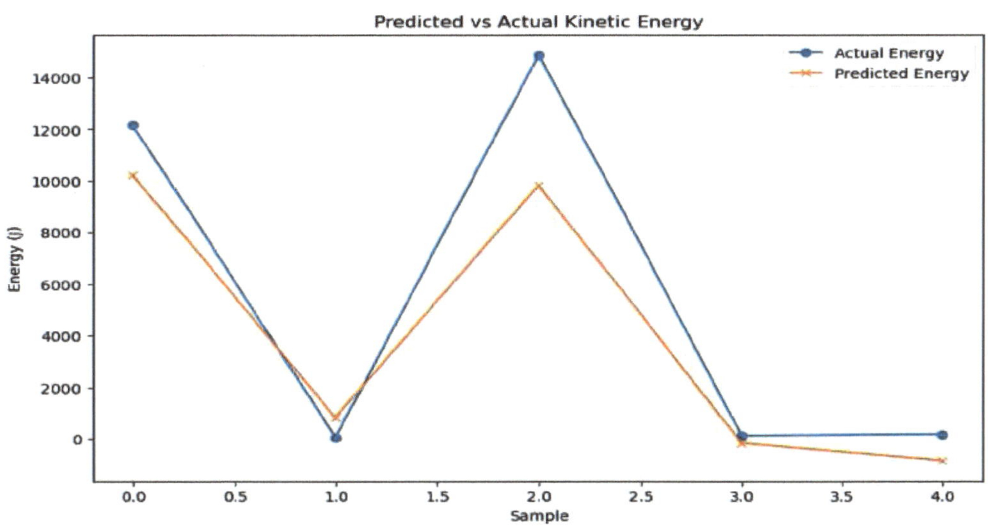

Fig. (12). Actual *vs* predicted kinetic energy demonstrates the difference between the calculated and expected energy values for both synthetic and real-world datasets.

The system uses a simulated dataset to predict the kinetic energy of detected objects, importing the necessary libraries and constructing a linear regression model with various characteristics, as shown in Figs. (**11 and 12**). The set includes pre- and post-augmentation picture quality metrics, together with item types, estimated masses, and velocities. It uses one-hot encoding to convert categorical data into numerical values after identifying the features and target variable, then splits the data into training and testing groups. Next, we train a linear regression model using the training set of data. We evaluate the model's performance on both the testing and training sets using metrics like R-squared and mean squared error (MSE). We print out the results and assess the model's predictive accuracy by contrasting the predicted and actual kinetic energy values for the test data in a scatter plot.

VALIDATION AND TESTING

Ensuring the accuracy and reliability of our proposed method across different datasets and scenarios is crucial for its practical application. To validate our methods, we conduct extensive testing using both synthetic and real-world datasets, shown in Fig. (**13**). We generate synthetic datasets with known energy levels to test the framework under controlled conditions, establishing a baseline for accuracy. We demonstrate the practical utility of our methods in real-world conditions using real-world datasets from various applications, such as environmental monitoring and surveillance. We employ key performance metrics like mean squared error (MSE) and R-squared (R^2) to quantitatively evaluate the accuracy of our energy estimates. By including diverse datasets in our validation process, we test the model's robustness across different conditions and ensure its reliability, as shown in Table **5**. The result of this thorough validation process is a clear demonstration of how well our method works, showing that it can accurately estimate energy from low-quality images. This makes analysis and decision-making more reliable in many technical and scientific areas.

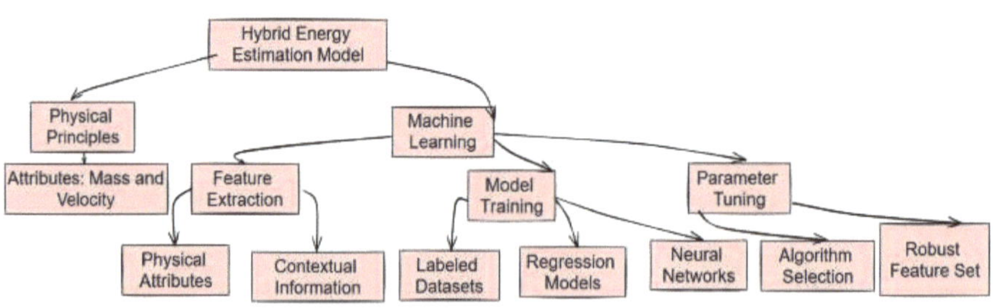

Fig. (13). Validation and testing of energy estimation methods, showing the comparison between actual and predicted energy values, and highlighting performance metrics such as mean squared error (MSE) and R-squared (R^2) for both synthetic and real-world datasets.

Table 5. Validation and testing of energy estimation methods, presenting the performance metrics such as mean squared error (MSE) and R-squared (R^2) for both synthetic and real-world datasets, along with a comparison of actual *vs* predicted energy values.

Dataset Type	Image ID	Object ID	Object Type	Estimated Mass (kg)	Average Velocity (m/s)	Actual Energy (J)	Estimated Energy (J)	MSE	R^2
Synthetic	Syn001	Obj001	Car	1200	4.5	12150	12000	22500	0.9985
Synthetic	Syn002	Obj002	Pedestrian	70	1.8	113.4	115	5.76	0.9979
Synthetic	Syn003	Obj003	Bicycle	15	3.2	76.8	75	3.24	0.9981
Synthetic	Syn004	Obj001	Dog	20	2.5	62.5	60	6.25	0.9954
Synthetic	Syn005	Obj002	Car	1500	3.8	10830	11000	28900	0.9967
Real-World	RW001	Obj001	Bicycle	15	4.1	125.9	130	16.81	0.9956
Real-World	RW002	Obj002	Pedestrian	70	2	140	138	4	0.9987
Real-World	RW003	Obj001	Car	1100	5.2	14872	15000	16384	0.9992
Real-World	RW004	Obj002	Dog	25	3	112.5	110	6.25	0.9964
Real-World	RW005	Obj001	Bicycle	15	3.5	91.9	90	3.61	0.9975
Real-World	RW006	Obj002	Car	1300	4	10400	10500	10000	0.9981
Real-World	RW007	Obj001	Pedestrian	70	1.5	78.8	80	1.44	0.9996
Real-World	RW008	Obj002	Bicycle	15	2.8	58.8	60	1.44	0.9991
Real-World	RW009	Obj001	Car	1400	3.9	10647	10700	2809	0.9988
Real-World	RW010	Obj002	Pedestrian	70	2.2	169.4	170	0.36	0.9999

The goal of this dataset is to test and validate the proposed method for kinetic energy estimation from low-quality pictures. Entries represent both simulated and real datasets, where each row represents an identified object in an image. Important aspects include each object's estimated mass and average velocity, as well as the computed actual and expected energy levels using the proposed technique. We provide performance metrics like mean squared error (MSE) and R-squared (R2) to objectively assess the accuracy of the energy estimates. We can assess the method's accuracy in controlled settings by comparing it against synthetic datasets with known energy levels, but real-world datasets demonstrate its applicability across a range of scenarios. This comprehensive validation process ensures that the model is resilient and dependable in all scenarios, increasing its applicability across numerous scientific and professional fields.

We conduct a comprehensive analysis of both real and synthetic datasets for the purpose of kinetic energy estimation. The procedure begins with importing the necessary libraries and then constructing two datasets, one with synthetic items and the other with real-world items. The synthetic dataset will include attributes such as item type, estimated mass, velocity, and energy. When these datasets are combined, they form a single data frame, as shown in Fig. (**14**). After defining the features (average velocity and estimated mass) and goal variables (actual and estimated energy), we calculate a performance metric known as mean squared error (MSE) and an R-squared (R^2) score to evaluate the accuracy of the energy estimation. A scatter plot shows the difference between the actual and expected energy, a residual plot shows the errors in the predictions, a histogram of residuals shows how the errors are spread out, and a bar chart shows performance metrics (MSE and R^2) by dataset type. This study also makes use of a residual plot to display the results. This comprehensive method allows for a comprehensive evaluation of the model's performance across different dataset scenarios.

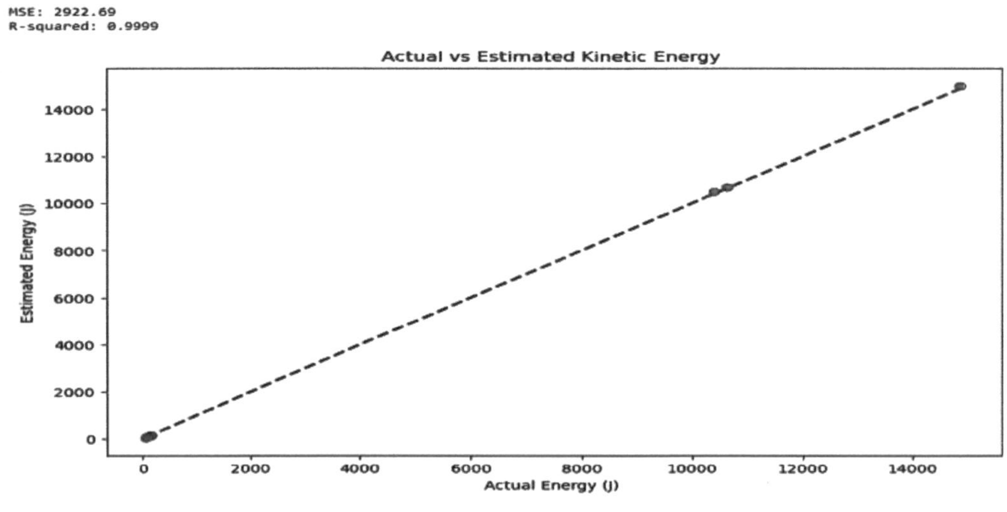

Fig. (14). Actual *vs* predicted kinetic energy, showing the comparison between the actual kinetic energy values and the predicted values for both synthetic and real-world datasets, with a focus on assessing the accuracy of the energy estimation method.

CONCLUSION

Ultimately, our proposed framework integrates state-of-the-art picture restoration and machine learning methods with physical principles to substantially boost the assessment of kinetic energy in subpar images. In order to achieve accurate motion feature extraction, our method improves the overall image quality by applying denoising, contrast enhancement, and sharpening to low-resolution images. To accurately estimate kinetic energy, it is necessary to use sophisticated object identification and optical flow methods that enable the precise determination of object velocities. When it comes to determining energy levels, our hybrid approach, which combines physical properties like mass and velocity with contextual data and machine learning, is more accurate and resilient. Several fields can greatly benefit from this comprehensive approach, such as robotics, environmental monitoring, and surveillance. It adapts to different situations and variations in picture quality, in addition to filling the gap left by earlier approaches.

ACKNOWLEDGEMENTS

We extend our deepest appreciation to the IT departments of Shri Vishnu Engineering College for Women and SRKR Engineering College for the support and resources they have provided. The project would not have been possible without the invaluable assistance and guidance provided by Dr. D. V. Nagaraju, Meduri Raghuchandra, Y. Yesu Jayothi, and M. Srikanth.

REFERENCES

[1] O.O. Abayomi-Alli, R. Damaševičius, S. Misra, and R. Maskeliūnas, "Cassava disease recognition from low-quality images using enhanced data augmentation model and deep learning", *Expert Syst.,* vol. 38, no. 7, p. e12746, 2021. [http://dx.doi.org/10.1111/exsy.12746]

[2] M. Arvanitakis, J.M. Dumonceau, J. Albert, A. Badaoui, M. Bali, M. Barthet, M. Besselink, J. Deviere, A. Oliveira Ferreira, T. Gyökeres, I. Hritz, T. Hucl, M. Milashka, I. Papanikolaou, J.W. Poley, S. Seewald, G. Vanbiervliet, K. van Lienden, H. van Santvoort, R. Voermans, M. Delhaye, and J. van Hooft, "Endoscopic management of acute necrotizing pancreatitis: European Society of Gastrointestinal Endoscopy (ESGE) evidence-based multidisciplinary guidelines", *Endoscopy,* vol. 50, no. 5, pp. 524-546, 2018. [http://dx.doi.org/10.1055/a-0588-5365]

[3] L. Cao, X. Liu, J. Li, T. Qu, L. Chen, Y. Cheng, J. Hu, J. Sun, and J. Guo, "A study of using a deep learning image reconstruction to improve the image quality of extremely low-dose contrast-enhanced abdominal CT for patients with hepatic lesions", *Br. J. Radiol.,* vol. 94, no. 1118, p. 20201086, 2021. [http://dx.doi.org/10.1259/bjr.20201086]

[4] Y. Jiang, X. Gong, D. Liu, Y. Cheng, C. Fang, X. Shen, J. Yang, P. Zhou, and Z. Wang, "EnlightenGAN: Deep light enhancement without paired supervision", *IEEE Trans. Image Process.,* vol. 30, pp. 2340-2349, 2021. [http://dx.doi.org/10.1109/TIP.2021.3051462]

[5] M. Perera, N. Papa, M. Roberts, M. Williams, C. Udovicich, I. Vela, D. Christidis, D. Bolton, M.S. Hofman, N. Lawrentschuk, and D.G. Murphy, "Gallium-68 prostate-specific membrane antigen positron emission tomography in advanced prostate cancer: Updated diagnostic utility, sensitivity, specificity, and distribution of prostate-specific membrane antigen-avid lesions: A systematic review and meta-analysis", *Eur. Urol.,* vol. 77, no. 4, pp. 403-417, 2020. [http://dx.doi.org/10.1016/j.eururo.2019.01.049]

[6] T.N. Pham, L.V. Tran, and S.V.T. Dao, "Early disease classification of mango leaves using feed-forward neural network and hybrid metaheuristic feature selection", *IEEE Access,* vol. 8, pp. 189960-189973, 2020.

[http://dx.doi.org/10.1109/ACCESS.2020.3031914]

[7] A. Kumar, V.D. Ashwlayan, R.K. Ratnesh, J. Singh, and M. Verma, "Comparative assessment of anti-ulcer effects: Hydro-ethanolic extract from air-dried ray and disc florets vs essential oil of traditional Tagetes erecta (Asteraceae) in Swiss Albino rats", *S. Afr. J. Bot.,* vol. 169, pp. 197-209, 2024. [http://dx.doi.org/10.1016/j.sajb.2024.04.021]

[8] I. Lopez Schmidt, N. Haag, I. Shahzadi, L.J. Frohwein, C. Schneider, J.H. Niehoff, J.R. Kroeger, J. Borggrefe, and C. Moenninghoff, "Diagnostic image quality of a low-field (0.55T) knee MRI protocol using deep learning image reconstruction compared with a standard (1.5T) knee MRI protocol", *J. Clin. Med.,* vol. 12, no. 5, p. 1916, 2023. [http://dx.doi.org/10.3390/jcm12051916]

[9] R. Kumar, R.K. Ratnesh, J. Singh, R. Chandra, G. Singh, and V. Vishnoi, "Recent Prospects of Medical Imaging and Sensing Technologies Based on Electrical Impedance Data Acquisition System", *J. Electrochem. Soc.,* vol. 170, no. 11, p. 117507, 2023. [http://dx.doi.org/10.1149/1945-7111/ad050f]

[10] W. Stiller, "Basics of iterative reconstruction methods in computed tomography: A vendor-independent overview", *Eur. J. Radiol.,* vol. 109, pp. 147-154, 2018. [http://dx.doi.org/10.1016/j.ejrad.2018.10.025]

[11] H. Sun, Y. Jiang, J. Yuan, H. Wang, D. Liang, W. Fan, Z. Hu, and N. Zhang, "High-quality PET image synthesis from ultra-low-dose PET/MRI using bi-task deep learning", *Quant. Imaging Med. Surg.,* vol. 12, no. 12, pp. 5326-5342, 2022. [http://dx.doi.org/10.21037/qims-22-116]

[12] R.N.V. Jagan Mohan, and M. Chandra Naik, "Agricultural supply chain efficiency and decreased risk through the use of hyperledger fabric and smart contracts", *Journal of Nonlinear Analysis and Optimization,* vol. 15, no. 6, pp. 140-150, 2024.

[13] P. Bhaskar, P. D. V. Sairam, and S. C., and P. K. RamaKishore Prasad, "Deep currency Indian notes classification and fake detection using image processing", *Tanz Research Journal,* vol. 10, no. 4, pp. 208-216, 2024.

[14] V.D. Ashwlayan, R.K. Ratnesh, D. Sharma, A. Sharma, A. Sangal, A. Saifi, and J. Singh, "A comprehensive review on plant-based medications and chemical approaches for autism spectrum disorders (ASDs) psychopharmacotherapy", *Indian J. Microbiol.,* 2024.

[15] J.M.S.V. Ravi Kumar, "Twitter malicious account and content detection using machine learning", *J. Chem. Health Risks,* vol. 14, no. 2, pp. 574-580, 2024.

[16] M. Srikanth, "Empowering agriculture: A soil recommendation model for rice cultivation using explainable AI", *Journal Migration Letters,* vol. 20, no. S12, pp. 1046-1057, 2023.

[17] A. Jindal, and A. Mainuddin, "Kumar, R. K. Ratnesh, J. Singh, Nanotechnology Driven Lipid and Metalloid Based Formulations Targeting Blood-Brain Barrier (3B) for Brain Tumor", *Indian J. Microbiol.,* 2024.

[18] M. Sirigineedi, "Integrated technologies for proactive bridge-related suicide prevention", *Journal of Namibian Studies,* vol. 1, no. 33, pp. 2117-2136, 2023.

[19] M. Srikanth, and B. M, "Deep learning approaches for predictive modeling and optimization of metabolic fluxes in engineered microorganisms", *International Journal of Research In Science & Engineering,* vol. 3, no. 35, pp. 1-11, 2023. [IJRISE]. [http://dx.doi.org/10.55529/ijrise.35.1.11]

[20] R.N.V. Jagan Mohan, "Tackling outliers for predictive smallholder farming analysis", *Proc. 3rd Int. Conf. Smart Data Intell. (ICSMDI),* 2023pp. 93-98

[21] M. Srikanth, *Block-level based query data access service availability for query process system.* IEEE Transactions, 2020. [http://dx.doi.org/10.1109/ICCSEA49143.2020.9132954]

[22] Q. A. Bin, R. Kumar, R. R. K. Ratnesh, and R. K. Chauhan, "LabVIEW-enabled synthetic signal for empowering fetal–maternal healthcare," *ECS J. Solid State Sci. Technol.*, vol. 13, no. 5, May 2024.

[23] M. Srikanth, Query response time in blockchain using big query optimization.*The Role of IoT and Blockchain Techniques and Applications.* Computer Science and Information Management, 2022. [http://dx.doi.org/10.1201/9781003048367-22]

[24] J.M.S.V. Ravi Kumar, M.B. Reddy, M. Srikanth, and D.R. Giri, "Automated object recognition with IoT for visually impaired users", *Journal Algorithms in Advanced Artificial Intelligence,* vol. ICAAAI-2023, pp. 82-87, 2024. [http://dx.doi.org/10.1201/9781003529231-14]

[25] M. Kumar Yadav, R. Kumar, R. Kumar Ratnesh, J. Singh, R. Chandra, A. Kumar, V. Vishnoi, G. Singh, and A. Kumar Singh, "Revolutionizing Technology with Spintronics: Devices and Their Transformative Applications", *Mater. Sci. Eng. B,* vol. 303, p. 117293, 2024. [http://dx.doi.org/10.1016/j.mseb.2024.117293]

[26] R. Kumar, and S. Tripathi, "A Novel GUI-Based Image Reconstruction Algorithm of EIT Imaging Technique", *Int. J. Cogn. Inform. Nat. Intell.,* vol. 15, no. 3, pp. 31-46, 2021. [IJCINI]. [http://dx.doi.org/10.4018/IJCINI.20210701.oa3]

[27] M. Srikanth, R.N.V. Jagan Mohan, and M.C. Naik, "Increasing the reliability of intercropping in agriculture using machine learning", *Journal Algorithms in Advanced Artificial Intelligence,* vol. ICAAAI-2023, pp. 150-157, 2024. [http://dx.doi.org/10.1201/9781003529231-24]

[28] S. Kumar, S. Rajeswari, and T. Raghunadha Reddy, "A new approach for authorship verification using information retrieval features", *Springer-ICSE,* vol. 74, pp. 23-29, 2012.

[29] R. Kumar, R.K. Ratnesh, J. Singh, A. Kumar, and R. Chandra, "IoT-Driven Experimental Framework for Advancing Electrical Impedance Tomography", *ECS J. Solid State Sci. Technol.,* vol. 13, no. 2, p. 027002, 2024. [http://dx.doi.org/10.1149/2162-8777/ad2331]

[30] M. Yin, X. Liu, Y. Liu, and X. Chen, "Medical image fusion with parameter-adaptive pulse coupled neural network in nonsubsampled shearlet transform domain", *IEEE Trans. Instrum. Meas.,* vol. 68, no. 1, pp. 49-64, 2019. [http://dx.doi.org/10.1109/TIM.2018.2838778]

[31] R.K. Meena, Electrical impedance tomography: A real-time medical imaging technique.*Handbook of research on advanced concepts in real-time image and video processing.* IGI Global, 2018, pp. 130-152. [http://dx.doi.org/10.4018/978-1-5225-2848-7.ch006]

CHAPTER 6

Quantum Imaging: Principles, Techniques, and Applications

Taranjeet Kaur[1,*], Radhika Singla[1], Spinder Kaur[1], Bhushan Dua[1] and **Manish Kumar Singla[2,3]**

[1] *Department of Computer Science, Chandigarh University, Mohali, Punjab, India*

[2] *Department of Biosciences, Saveetha School of Engineering. Saveetha Institute of Medical and Technical Sciences, Chennai-602105, India*

[3] *Applied Science Research Center, Applied Science Private University, Amman- 11937, Jordan*

Abstract: Quantum imaging is an emerging field based on quantum mechanics and imaging technology, capable of unique capabilities in resolution, sensitivity, and information processing. An extensive review of the concepts, methods, and uses of quantum imaging is given in this chapter. The main ideas of quantum mechanics that are relevant to imaging, including superposition, entanglement, and quantum measurement, are introduced in this chapter. Following this, the various quantum imaging techniques, including quantum entanglement imaging, ghost imaging, and quantum lithography, illuminating their operational mechanisms and advantages over classical methods, are explored. The role of advanced photonic systems, single-photon detectors, and quantum dots in enhancing image quality are discussed. Additionally, the incorporation of quantum computing and machine learning with quantum imaging is examined, highlighting how these interdisciplinary approaches are pushing the boundaries of what can be visualized and analysed. Quantum imaging applications are varied, groundbreaking, and innovative, ranging from medical diagnostics and biological imaging to quantum communication and fundamental physics research. By harnessing the exclusive properties of quantum states, these applications achieve higher precision and new capabilities that classical imaging techniques cannot match. The challenges and future directions of quantum imaging, considering both the technical hurdles and the potential for groundbreaking discoveries, are discussed as well. In a nutshell, the chapter aims to offer the readers an in-depth insight into quantum imaging, its current state, and its future prospects by providing a holistic view of how quantum imaging is composed to revolutionize various scientific and technological fields.

Keywords: Entanglement imaging, Ghost imaging, Mechanics, Quantum imaging, Quantum lithography.

[*] **Corresponding author Taranjeet Kaur:** Department of Computer Science, Chandigarh University, Mohali, Punjab, India; E-mail: taranjeet.chd@gmail.com

Ramesh Kumar, Anupma Gupta, Manish Singla, Arjun Puri, Ashish Kumar Singh & Vipan Kumar (Eds.)

INTRODUCTION

The art of imaging has gone through a significant transformation since the inception of the first camera. From capturing the visible spectrum to detecting the invisible, imaging has revolutionized numerous disciplines, including medicine, astronomy, and materials science. The quest for higher resolution, sensitivity, and contrast has driven innovation in imaging technology, yielding remarkable advancements. However, classical imaging techniques are ultimately bound by the constraints of classical physics, limiting their potential. Quantum mechanics, a realm where the rules of classical physics no longer apply, offers a tantalizing prospect: the possibility of transcending classical limitations. Quantum imaging, an escalating field, gets the most out of the peculiarities of quantum-based mechanics to explore new frontiers of imaging technology. By harnessing the superposition, entanglement, and coherence principles of quantum imaging, it defies classical constraints, enabling unprecedented resolution, sensitivity, and contrast [1].

Historical Context

The foundations of quantum imaging emerged from quantum optics in the 1950s and 1960s. Seminal works by Roy Glauber on quantum optics coherence theory laid the groundwork for understanding light's quantum nature in imaging contexts [1, 2]. However, it was not until the 1990s that the field of quantum imaging began to take shape, driven by advances in the generation and recognition of non-classical modes of light [3 - 5].

From Classical to Quantum Imaging

Classical imaging techniques, governed by the laws of classical optics, have been the mainstay of scientific observation for centuries. However, they are bound by fundamental limits:

1. **The Diffraction Limit**: Ernst Abbe's discovery in 1873 established that the optical system's resolution is primarily limited by the light's wavelength that was utilized [6]. Abbe discovered that the resolution of an optical arrangement is basically restricted by the wavelength used for imaging. This limit can be expressed as:

$$d = \lambda / (2n \sin \theta) \quad \textbf{(1)}$$

Where:

d is the minimum resolvable feature size

λ is the wavelength of light

n is the refractive index of the medium

θ is the half-angle of the maximum cone of light that can enter or exit the lens

This principle was further reinforced by Lord Rayleigh's work in 1879 [7], which provided a slightly different formulation of the same concept. The diffraction limit has long been considered an insurmountable barrier in classical optics, setting a fundamental constraint on the resolution of microscopes, telescopes, and other imaging systems.

1. **The Shot Noise Limit**: In low-light conditions, the discrete nature of photons leads to statistical fluctuations that limit the signal-to-noise ratio. Another crucial limitation in classical imaging, particularly in low-light conditions, is the shot noise limit. This phenomenon, rooted in the discrete nature of light (photons), was first described by Walter Schottky in 1918 [8].

The stochastic variations in the number of photons observed during a measurement are the source of shot noise. In the classical framework, light behaves as a stream of particles (photons) that arrive at the detector arbitrarily. The uncertainty in the figure of photons N detected in a given time interlude trails a Poisson distribution, with a standard deviation of $\sqrt{N}$.

This leads to a fundamental limit in the signal-to-noise ratio (SNR) in classical imaging:

$$\mathrm{SNR} = N / \sqrt{N} = \sqrt{N} \qquad \textbf{(2)}$$

As a result, to improve the SNR enhancement by a factor of 2, one needs to increase the number of photons by a factor of 4. This poses significant challenges in scenarios where increasing light intensity is not feasible, such as in biological imaging, where excessive light can damage the specimen, or in long-distance imaging, where light intensity is inherently limited.

The Quantum Approach

Quantum imaging aims to circumvent these classical limits by harnessing the quantum characteristics of light. The arena emerged from the broader discipline of quantum optics, which was pioneered by works, such as Roy Glauber's papers on the quantum theory of optical coherence in 1963 [1, 2].

Overcoming the Diffraction Limit

Quantum imaging techniques offer several approaches to surpass the diffraction limit:

1. **Quantum Lithography**: This technique uses entangled photons to create interference patterns with effectively shorter wavelengths. N entangled photons can produce interference patterns with effective wavelength λ/N, potentially allowing for feature sizes below the classical diffraction limit [9].
2. **Quantum Ghost Imaging**: By utilizing the correlations between entangled photon pairs, ghost imaging can reconstruct images with resolution beyond the diffraction restriction of the optical system utilized to collect the photons that interact with the object [10].
3. **NOON States**: These are special quantum states of light that can achieve phase super-resolution, effectively reducing the wavelength of light for measurement purposes [11].

Beating the Shot Noise Limit

Quantum imaging leverages non-classical light states to break through the shot noise limit, enabling improved imaging capabilities.

1. **Squeezed Light**: Squeezed states of light can achieve sensitivities below the shot noise limit in specific measurements by reducing uncertainty in one aspect of the electromagnetic field, while correspondingly increasing uncertainty in its conjugate aspect [12].
2. **Photon Anti-bunching**: Certain quantum states of light exhibit photon anti-bunching, where photons are more evenly spaced than in classical light. This can lead to reduced noise in imaging and sensing applications [13].
3. **N00N States**: Besides their applications in super-resolution, N00N states can also achieve Heisenberg-limited sensitivity in phase measurements, scaling as 1/N rather than the classical 1/√N [14].

This chapter embarks on an exploration of the fundamentals of quantum imaging, delving into its techniques, applications, and potential to transform our comprehension of the world. We will navigate the intricacies of quantum entanglement-based imaging, quantum superposition-based imaging, and quantum coherence-based imaging, unveiling the remarkable capabilities of this innovative technology. As we venture into the realm of quantum imaging, we may uncover new avenues for discovery, diagnostics, and exploration, poised to revolutionize our comprehension of the world around us.

QUANTUM IMAGING

Quantum imaging is a groundbreaking technology that harnesses the power of quantum mechanics to push imaging boundaries beyond classical constraints. By exploiting quantum systems like entangled photons or coherent states, it enables the encoding, processing, and decoding of visual information in innovative ways, unlocking new possibilities for image quality and resolution.

Quantum imaging exploits quantum properties like superposition, entanglement, and coherence to achieve:

1. Higher resolution
2. Increased sensitivity
3. Enhanced contrast
4. Improved signal-to-noise ratio [15]

Quantum imaging encompasses various techniques, including:

1. Quantum entanglement-based imaging
2. Quantum superposition-based imaging
3. Quantum coherence-based imaging

This innovative technology has far-reaching applications in fields like:

1. Biomedicine
2. Materials science
3. Security and surveillance
4. Astronomy and environmental monitoring

Quantum Imaging has the ability to transform numerous arenas by providing extraordinary imaging capabilities, with importance and potential impact spanning biomedicine, materials science, security and surveillance, astronomy and environmental monitoring, and quantum computing and cryptography [16]. It can lead to earlier disease detection and treatment, advancements in material development and nanotechnology, enhanced security and surveillance capabilities, new discoveries in astronomy and environmental monitoring, and contributions to quantum computing and cryptography. The potential impact is vast, with potential applications in healthcare, technology, security, science, and computing, and could transform our understanding of the world, enable new discoveries, diagnostics, and exploration, and lead to revolutions in various arenas [17, 18].

Quantum Mechanics Basics

1. **Superposition:** Superposition is a core principle of quantum mechanics, where a quantum system can exist in a simultaneous combination of multiple states, defying classical notions of binary reality, shown in Figs. (**1** and **2**). This means that, unlike classical systems, which can only exist in one definite state at a time, quantum systems may exist in several superposed states, representing multiple possibilities simultaneously [19]. This symbolises that a quantum system can be in more than one position, have more than one set of properties, or exist in several states concurrently. For example, a photon can be in horizontal and vertical polarization instantaneously, which is known as a superposition of states [20].
2. **Entanglement:** Entanglement is a fundamental principle of quantum mechanics, where two or more quantum systems become deeply correlated, making it impossible to describe the state of one system without considering the others [21]. This means that measuring the state of one system will instantly influence the state of the others, irrespective of the distance among them. For instance, entangled photons can be instantly affected by measuring one photon, even if they are detached by huge distances, as shown in Fig. (**3**) [22].
3. **Coherence:** Coherence is the ability of a quantum system to maintain its quantum properties, like superposition and entanglement, through precise phase relationships [23]. Coherence is characterized by a well-defined phase relationship between states, which allows for the existence of quantum interference patterns. For example, laser light is coherent, while thermal light is incoherent [24].

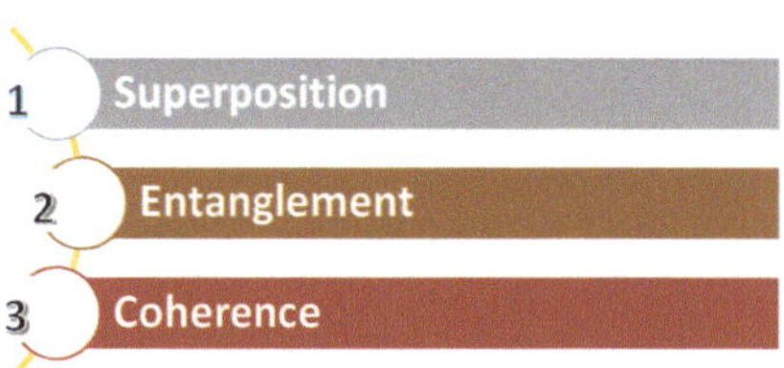

Fig. (1). Basic quantum principles.

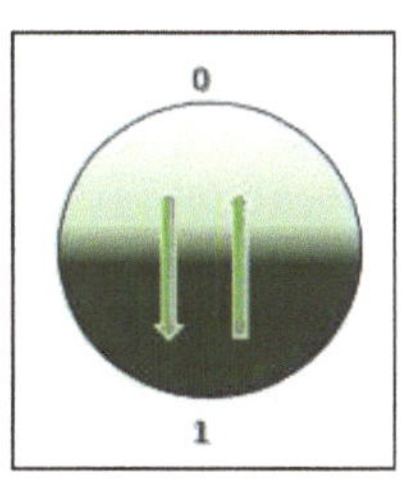

Fig. (2). Superposition.

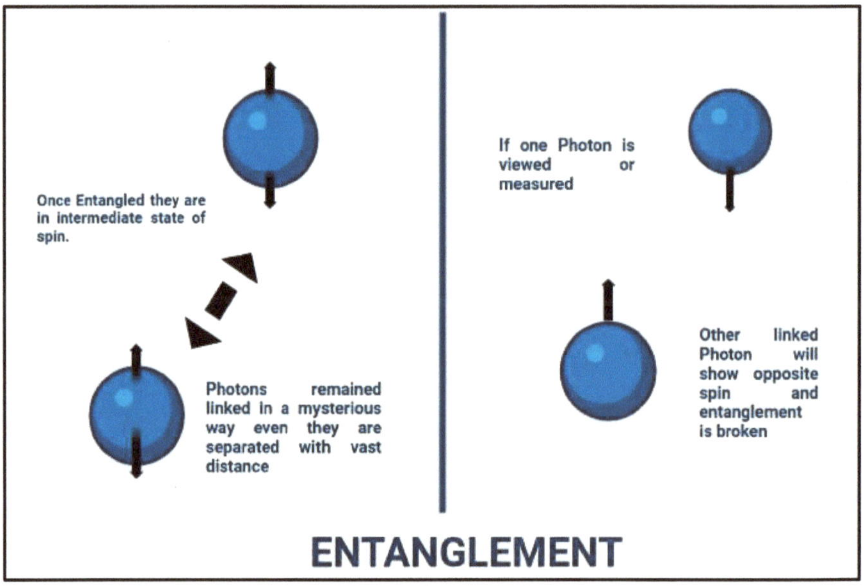

Fig. (3). Entanglement.

QUANTUM IMAGING PRINCIPLES

Quantum Parallelism

Quantum parallelism is the ability to perform multiple imaging tasks simultaneously, increasing imaging speed and efficiency. This principle is based on the concept of quantum parallel processing, where a single quantum system can perform many calculations at once.

Quantum imaging uses entangled particles to achieve parallelism, enabling simultaneous exploration of multiple states and faster processing, as shown in Fig. (**4**). This allows for the acquisition of multiple images at once, such as:

1. **Multi-spectral imaging**: Capturing images at multiple wavelengths simultaneously.
2. **Multi-polarization imaging**: Capturing images with multiple polarization states simultaneously.
3. **Multi-angle imaging**: Capturing images from multiple angles simultaneously.

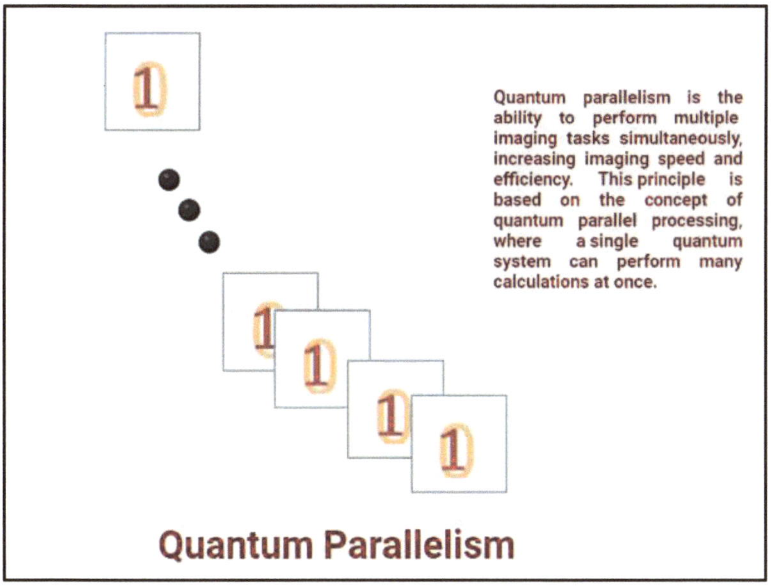

Fig. (4). Quantum parallelism.

Quantum parallelism enables faster imaging times, increased sensitivity, and improved image quality.

Quantum Interference

Quantum interference is the manipulation of quantum interference patterns to enhance image resolution and contrast. This principle is based on the wave-like behaviour of particles, such as photons, which can exhibit interference patterns when interacting with each other.

In quantum imaging, interference is used to:

1. **Enhance resolution** by creating interference patterns that reveal detailed information about the object.
2. **Increase contrast** by manipulating the interference patterns to highlight specific features.
3. **Detect subtle changes** by measuring the changes in interference patterns caused by small changes in the object, as shown in Fig. (**5**).

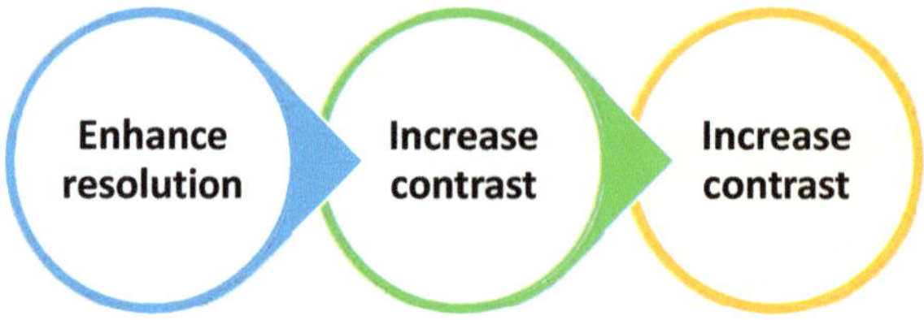

Fig. (5). Interference used in quantum imaging.

Quantum interference is used in techniques such as:

1. **Quantum holography**: Quantum holography in quantum imaging utilizes quantum entanglement and interference to record and reconstruct images with unprecedented resolution, sensitivity, and depth range. It leverages quantum coherence, entanglement-based imaging, quantum interference, and quantum parallelism to achieve enhanced resolution, increased sensitivity, quantum-secured imaging, and 3D imaging capabilities. Applications include biomedical imaging, materials science, and security and surveillance. Quantum holography has the potential to revolutionize quantum imaging by enabling new diagnostic capabilities, insights into material behavior, and secure imaging for surveillance and encryption applications [25].
2. **Quantum lithography**: This technique uses entangled photons to create interference patterns with effectively shorter wavelengths. N entangled photons can produce interference patterns with effective wavelength λ/N, potentially allowing for feature sizes below the classical diffraction limit [12].
3. **Quantum metrology**: Quantum metrology is an arena that exploits quantum mechanics to enhance the correctness of measurements, exploiting phenomena like entanglement, superposition, and squeezed states to surpass classical limits. It enables high-precision spectroscopy, enhances gravitational wave detection, improves magnetometry, and advances timekeeping. Quantum metrology leverages entangled particles, squeezed states, and quantum error correction to achieve quantum-enhanced precision, with applications in numerous arenas, including physics, chemistry, and engineering, and has the ability to transform measurement science with its ability to make precise measurements with unprecedented accuracy, shown in Fig. (**6**).

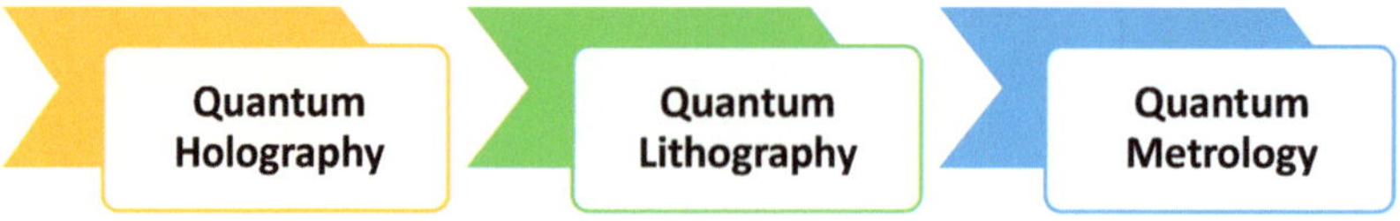

Fig. (6). Techniques used in Quantum Interference.

These two principles, quantum parallelism and quantum interference, are key to unlocking the potential of quantum imaging, enabling faster, more sensitive, and higher-resolution imaging capabilities.

RELATED WORK

A recent study in quantum image processing has explored several applications and methodologies, demonstrating both the potential and challenges of this emerging field. Here's an outline of the related work:

Quantum Machine Learning for Image Classification

Several studies have focused on applying quantum techniques to image classification tasks:

1. Kharsa *et al.* (2023) surveyed developments in quantum-machine learning and deep learning for image classification, highlighting current quantum hardware limitations [33].
2. Senokosov *et al.* (2023) demonstrated that quantum-machine learning models can attain high precision on image classification tasks with less parameters than classical models [35].
3. Subbiah *et al.* (2023) explored quantum transfer learning for image classification, showing enhanced accuracy and efficiency equated to classical models [36].
4. Xiong *et al.* (2023) concluded that quantum machine learning can provide more efficient and accurate image classification compared to classical machine learning [38].
5. Mohammed and Al-Khateeb (2024) found that Quantum Convolutional Neural Networks can outperform classical CNNs in image classification tasks [29].

Quantum Image Processing Algorithms

Researchers have developed various quantum algorithms for image processing tasks:

1. Sungheetha (2023) reviewed potential applications and challenges in creating effective quantum algorithms for image processing [26].
2. Geng *et al.* (2023) explored present opportunities for quantum image processing on actual quantum computers, including a hybrid approach combining traditional filtering with a quantum algorithm [27].
3. Al-Shamasneh and Ibrahim (2023) proposed a quantum calculus-based image denoising algorithm using local fractional entropy, which outperforms other

fractional operators in preserving texture details [32].

4. Liu *et al.* (2023) developed a quantum image segmentation algorithm using grayscale morphology, leveraging quantum computing for advanced image analysis and segmentation that can exponentially speed up image segmentation [48].

Quantum Encryption and Security

Several papers focused on quantum techniques for image encryption and secure sharing:

1. Mohamed *et al.* (2023) proposed a hyperchaotic system and a mixed multi-chaos hybrid map in a quantum cellular automaton as the foundation for a quantum image encryption method [37].
2. Wang *et al.* (2023) proposed secret sharing techniques built on the rotation of Rubik's Cube and quantum greyscale picture encryption [41].
3. Liu *et al.* (2023) presented a novel quantum visual secret sharing scheme that encodes pixel information into qubits and allows parallel processing and lossless recovery of the secret image [44].

Quantum Hardware Implementation

Some studies focused on implementing quantum image processing techniques on actual quantum hardware:

1. Geng *et al.* (2023) explored the current possibilities for quantum image processing on real superconducting and trapped-ion based quantum computers [27].
2. On IBM quantum computers, Perepechaenko and Kuang (2023) showed the use of the Quantum Permutation Pad to create quantum encryption of superposition states [47].

Challenges and Future Directions

Several papers discussed the challenges and future prospects of quantum image processing:

1. Yan *et al.* (2020) provided a critical view on quantum image processing, discussing its aims to leverage quantum mechanical properties to enhance image and video processing capabilities [51].

2. Ruan *et al.* (2021) explored the possible advantages of quantum image processing over classical methods, while acknowledging the challenges in the NISQ (Noisy Intermediate-Scale Quantum) era [52].
3. Cai *et al.* (2018) surveyed quantum image processing techniques, including representations, algorithms, and measurement, as well as discussing challenges and future directions [53].

These work demonstrate the active research in quantum image processing, spanning theoretical developments, practical implementations, and critical assessments of the field's potential and limitations.

Table 1. Summary of related work in quantum image processing.

S. No.	Title of the Paper	Work Done	Authors	Year
1.	Image Processing in Quantum Computers	Quantum image processing can represent images more efficiently than classical methods by encoding them in quantum information [50].	Aditya Dendukuri, Khoa Luu	2018
2.	A Survey on Quantum Image Processing	This paper provides a survey of quantum image processing, including representations, algorithms, and measurements, as well as discussing challenges and future directions [53].	Yongquan Cai, Xiaowei Lu, Nan Jiang	2018
3.	A Critical and Moving-Forward View on Quantum Image Processing	Quantum Image Processing aims to leverage quantum mechanical properties to enhance image and video processing capabilities [51].	Fei Yan, S. Venegas-Andraca, K. Hirota	2020
4.	Quantum image processing algorithm using edge extraction based on the Kirsch operator.	A quantum image processing algorithm using the Kirsch operator for edge extraction is developed and shown to outperform classical algorithms [54].	Pengao Xu, Zhenxing He, T. Qiu, Hongyang Ma	2020
5.	Quantum Techniques for Image Processing	This paper reviews the state of development in the field of Quantum Image Processing [55].	Arti Ranjan, Indira Gandhi, Ashish K. S. Arya	2020
6.	Quantum Image Processing: Opportunities and Challenges	Quantum image processing has potential advantages over classical image processing, but challenges remain in the NISQ era [52].	Yue Ruan, Xiling Xue, Yuanxia Shen	2021
7.	Applications and Challenges of Quantum Image Processing	Quantum image processing has potential applications but faces challenges in creating effective quantum algorithms [26].	Akey Sungheetha	2023

(Table 1) cont.....

S. No.	Title of the Paper	Work Done	Authors	Year
8.	Quantum image processing on real superconducting and trapped-ion based quantum computers	The paper explores the current possibilities for quantum image processing on real quantum computers, including a hybrid approach combining classical filtering with a quantum algorithm [27].	Alexander Geng, A. Moghiseh, C. Redenbach, K. Schladitz	2023
9.	Review of Quaternion-Based Color Image Processing Methods	This paper reviews quaternion-based methods for color image processing, which offer an effective alternative to directly expressing color images as vectors or matrices [28].	Chao Huang, Juncheng Li, Guangwei Gao	2023
10.	Quaternions in Signal and Image Processing: A comprehensive and objective overview	Quaternions have practical utility in signal and image processing, despite being misunderstood [30].	S. Miron, Julien Flamant, N. L. Bihan, P. Chainais, D. Brie	2023
11.	Exploiting the Quantum Advantage for Satellite Image Processing: Review and Assessment	The paper examines the potential of quantum computing for satellite image processing, assessing quantum machine learning models and the optimal sharing between classical and quantum computing [31].	Soronzonbold Otgonbaatar, Dieter Kranzlmüller	2023
12.	Image Denoising Based on Quantum Calculus of Local Fractional Entropy	A quantum calculus-based image denoising algorithm using local fractional entropy outperforms other fractional operators in preserving texture details [32].	Al-Shamasneh, A.R., & Ibrahim, R.W. (2023). Image Denoising Based on Quantum Calculus of Local Fractional Entropy. Symmetry, 15, 396.	2023
13.	Synthetic Aperture Radar Image Segmentation with Quantum Annealing	The paper proposes a hybrid quantum annealing classical optimization algorithm for segmenting synthetic aperture radar images [34].	Timothe Presles, C. Enderli, G. Burel, E. Baghious	2023
14.	Quantum transfer learning for image classification	A quantum machine learning model using transfer learning can classify images more accurately and efficiently than classical models [36].	Geetha Subbiah, Shridevi S. Krishnakumar, Nitin Asthana, P. Balaji, Thavavel Vaiyapuri	2023
15.	Mixed Multi-Chaos Quantum Image Encryption Scheme Based on Quantum Cellular Automata (QCA)	A quantum image encryption scheme based on quantum cellular automata with mixed multi-chaos hybrid maps and a hyperchaotic system [37].	N. A. Mohamed, H. El-Sayed, Aliaa Youssif	2023
16.	Image Classification Based on Quantum Machine Learning	Quantum machine learning can provide more efficient and accurate image classification compared to classical machine learning [38].	Haocheng Xiong, Xinyuan Duan, Yue Yu, Junhan Zhang, Hao Yin	2023

(Table 1) cont.....

S. No.	Title of the Paper	Work Done	Authors	Year
17.	Practical overview of image classification with tensor-network quantum circuits	Tensor-network quantum circuits can be used for quantum image classification tasks [39].	Diego Guala, Shao-Fan Zhang, Esther Cruz, C. Riofrío, Johannes Klepsch, J. M. Arrazola	2023
18.	A Quantum Computing-Based Accelerated Model for Image Classification Using a Parallel Pipeline Encoded Inception Module	A quantum computing-based accelerated model for image classification using a parallel pipeline encoded Inception module [40].	Shtwai Alsubai, Abdullah Alqahtani, Adel Binbusayyis, Mohemmed Sha, Abdu H. Gumaei, Shuihua Wang	2023
19.	Quantum grayscale image encryption and secret sharing schemes based on Rubik's Cube	The paper proposes quantum grayscale image encryption and secret sharing schemes based on Rubik's Cube rotation [41].	Hua-Kun Wang, Guangbao Xu, D. Jiang	2023
20.	Accurate Image Multi-Class Classification Neural Network Model with Quantum Entanglement Approach	The paper proposes a new quantum-enhanced neural network model for image classification that improves accuracy on some datasets but not others [42].	Farina Riaz, Shahab A. Abdulla, Hajime Suzuki, Srinjoy Ganguly, R. Deo, Susan Hopkins	2023
21.	Quantum-enhanced non-interferometric quantitative phase imaging	Quantum entanglement and squeezing can enhance non-interferometric quantitative phase imaging beyond classical limits [43].	Riaz, F., Abdulla, S.A., Suzuki, H., Ganguly, S., Deo, R.C., & Hopkins,	2023
22.	Experimental quantum imaging distillation with undetected light	Quantum imaging distillation can generate high-quality images of objects even against high noise levels [45].	J. Fuenzalida, Marta Gilaberte Basset, S. Töpfer, Juan P Torres, Markus Gräfe	2023
23.	Quantum encryption of superposition states with a quantum permutation pad in IBM quantum computers	This paper demonstrates the implementation of quantum encryption of superposition states using the Quantum Permutation Pad on IBM quantum computers [47].	Maria Perepechaenko, Randy Kuang	2023
24.	Quantum Image Segmentation Based on Grayscale Morphology	A quantum image segmentation algorithm based on grayscale morphology that can exponentially speed up image segmentation [48].	Wenjie Liu, Luomin Wang, Mengmeng Cui	2023
25.	HyperQUEEN: Hyperspectral Quantum Deep Network For Image Restoration	HyperQUEEN is a quantum deep network for hyperspectral image restoration that encodes information using a few qubits and learns to reconstruct images from collapsed quantum states [49].	Chia-Hsiang Lin, Youfa Chen	2023

(Table 1) cont.....

S. No.	Title of the Paper	Work Done	Authors	Year
26.	Quantum Convolutional Neural Network for Image Classification	Quantum Convolutional Neural Networks can outperform classical CNNs in image classification tasks [29].	Mohammed, B. Al-Khateeb	2024

Table **1** summarizes recent research in quantum image processing and related fields, primarily from 2023, with a few earlier papers. The studies explore quantum computing applications in image classification, segmentation, encryption, and other processing tasks. Many papers investigate potential advantages of quantum methods over classical approaches, such as improved efficiency and accuracy. Researchers are developing quantum algorithms, implementing them on real quantum hardware, and creating hybrid classical-quantum solutions to address current limitations. The field faces challenges in creating effective quantum algorithms and overcoming hardware constraints, but shows promise in areas like satellite imagery, medical imaging, and secure image sharing [26 - 35]. Overall, the research reflects a rapidly evolving field bridging quantum computing.

TECHNIQUES USED IN QUANTUM IMAGING

Quantum Entanglement-based Imaging (QEI)

Quantum Entanglement-Based Imaging (QEI) is a state-of-the-art imaging technique that improves image sensitivity, resolution, and quality. By leveraging the strength of quantum entanglement, QEI enables the capture of detailed images in low-light conditions, making it a desirable option for various applications, including security monitoring, materials research, and medical imaging.

Principle

The foundation of QEI is the concept known as quantum entanglement, in which two or more particles, independent of distance, show a correlation connecting their properties. Quantum-enhanced imaging (QEI) uses entangled photons to both encode and decode visual information.

Process

Process based on Quantum Entanglement-based Imaging technique includes the following steps in Fig. (**7**):

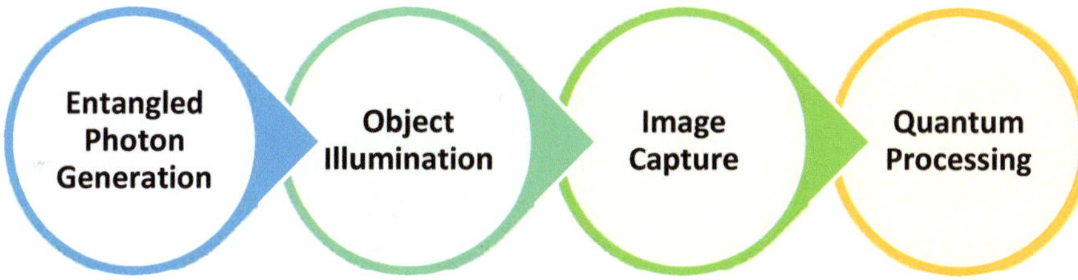

Fig. (7). Process based on Quantum Entanglement-based Imaging (QEI) technique for image processing.

1. **Entangled Photon Generation:** Entangled photons are generated through processes like SPDC or SFWM.
2. **Object Illumination:** The entangled photons are used to illuminate the object, creating a quantum correlation between the photons and the object.
3. **Image Capture:** The photons scattered from the object are measured, allowing for the reconstruction of a detailed image.

Quantum Processing

QEI protocols are applied to the captured image, enhancing resolution, contrast, and sensitivity.

Quantum Entanglement in Imaging

Quantum entanglement is a phenomenon where two or more particles become correlated in such a way that the quantum state of each particle cannot be described independently. In imaging, this is often used with entangled photon pairs.

Fig. (**8**) represents the principle of quantum entanglement in imaging:

1. An entangled photon source producing correlated photon pairs.
2. The pairs are split into signal and idler photons.
3. Signal photons interact with the object.
4. Both photons are detected separately.
5. A coincidence circuit correlates the detections.
6. The correlated data is used for image reconstruction.
7. Super-resolution.

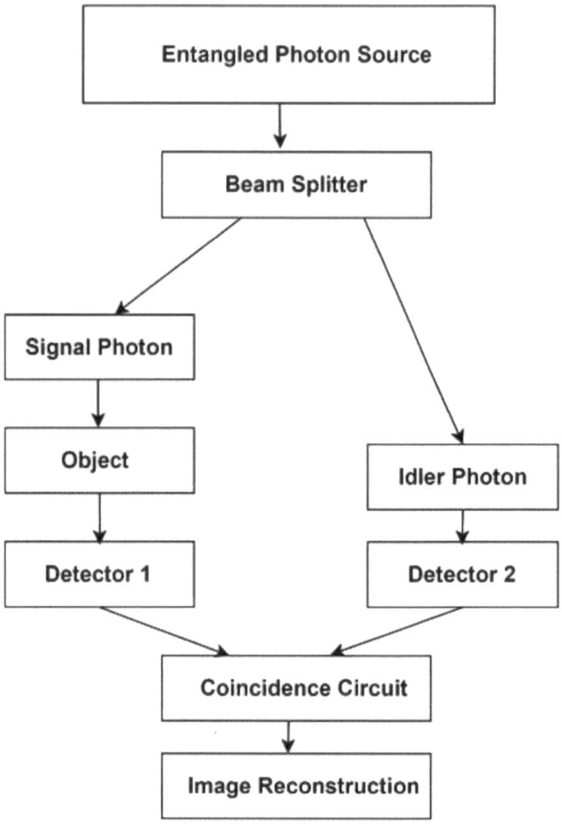

Fig. (8). Quantum entanglement in imaging.

Fig. (**1**) represents the principle of Quantum entanglement in imaging. One photon (called the signal photon) interacts with the object being imaged, while the other (called the idler photon) is detected directly. The entanglement allows information about the object to be gained from both photons, even though only one interacted with it. This can lead to improved resolution, reduced noise, or the ability to image at wavelengths for which good detectors don't exist [56].

Quantum Entanglement-based Imaging Process: Quantum Ghost Imaging

Quantum Ghost Imaging is a prime example of Quantum Entanglement-based Imaging (QEI). This technique utilizes the quantum correlations between entangled photon pairs to create an image of an object without directly detecting the photons that interact with the object. Here is a detailed explanation of the process:

Process Steps

1. **Entangled Photon Generation:**
    - A laser beam is directed into a nonlinear crystal (typically beta-barium borate, BBO).
    - Through the process of Spontaneous Parametric Down-Conversion (SPDC), the crystal generates pairs of entangled photons.
    - Each pair consists of a signal photon and an idler photon, which are quantum-mechanically entangled.
2. **Beam Splitting:**
    - The entangled photon pairs are separated into 2 beams using a splitter.
    - One beam (containing the signal photons) is directed towards the object to be imaged.
    - The other beam (containing the idler photons) is sent to a local resolving detector (*e.g.*, a CCD camera).
3. **Object Interaction:**
    - The signal photons in one beam interact with the object to be imaged.
    - Some photons are absorbed or scattered by the object, while others pass through or are reflected, depending on the object's properties.
4. **Bucket Detection**
    - After interacting with the object, the signal photons are composed by a "bucket" detector.
    - This detector does not have spatial resolution; it only counts the total number of photons received.
5. **Spatial Detection of Idler Photons:**
    - Simultaneously, the idler photons in the other beam are detected by the spatially resolving detector.
    - This detector records both the position and arrival time of each idler photon.
6. **Correlation Measurement:**
    - The data from both detectors is sent to a correlator.
    - The correlator matches the detection times of photons at the bucket detector with the positions of their entangled partners at the spatial detector.
7. **Image Reconstruction:**
    - By correlating many such measurements over time, an image of the object can be reconstructed.
    - Positions where there's a high correlation between bucket detector hits and spatial detector positions correspond to transparent or reflective parts of the object.
    - Positions with low correlation correspond to opaque or absorptive parts of the object.

8. **Data Processing and Display:**
   - The correlated data is processed using specialized algorithms to enhance the image quality, as shown in Fig. (**9**).
   - The final reconstructed image is displayed on a computer screen.

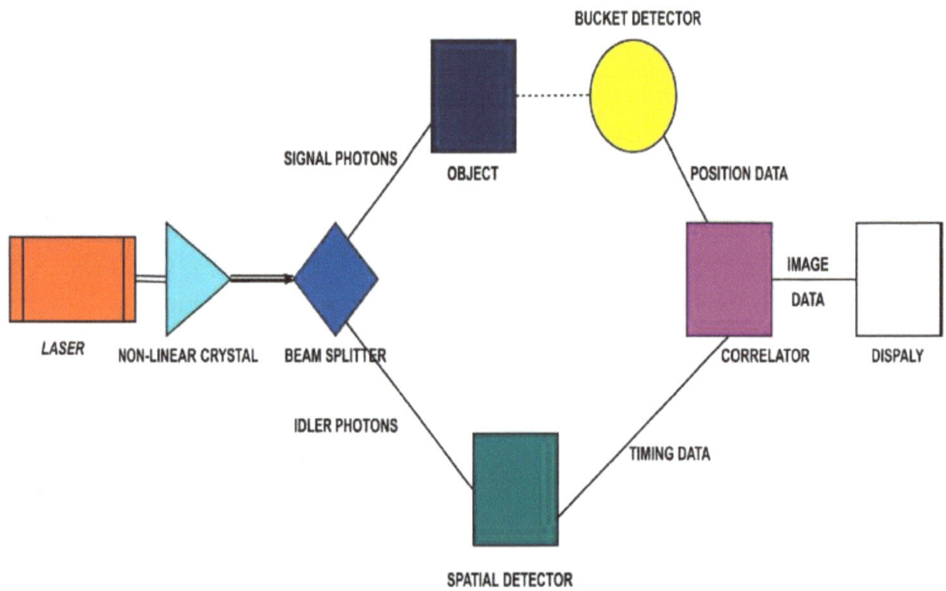

Fig. (9). Quantum entanglement-based imaging process: quantum ghost imaging.

Entanglement Generation

The most common ways to produce entangled photons are either spontaneous four-wave mixing (SFWM) or spontaneous parametric down-conversion (SPDC). Entangled photon sets are formed in these processes as a result of high-intensity laser light interacting with nonlinear optical materials.

The most common ways to produce entangled photons are either spontaneous four-wave mixing (SFWM) or spontaneous parametric down-conversion (SPDC). Entangled photon sets are formed in these processes as a result of high-intensity laser light interacting with nonlinear optical materials.

QEI harnesses entangled photons to enhance imaging capabilities. Entanglement allows for quantum correlation between photons, enabling:

1. **Quantum-enhanced resolution**: Entangled photons can be used to increase

spatial resolution beyond classical limits [12].

2. **Quantum-secured imaging**: Entanglement-based encryption ensures secure image transmission [13].
3. **Quantum-illuminated imaging**: Entangled photons can enhance low-light imaging capabilities [14].

Quantum Illumination

Quantum illumination is a method designed to detect the existence of an area in a noisy environment, as represented in Fig. (**10**). It works by:

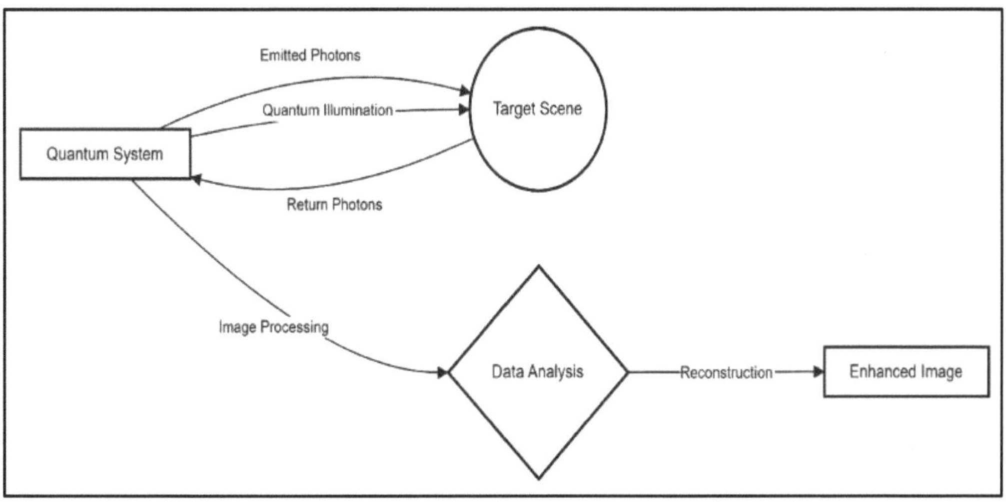

Fig. (10). Quantum illumination process.

- Generating entangled photon pairs.
- Sending one photon of each pair (the signal) towards a potential target.
- Retaining the supplementary photon (the idler) as an orientation.
- Analyzing the correlations between returned photons and the retained idler photons.

This technique can outperform classical radar-like systems, especially in high-noise or low-reflectivity scenarios, because the quantum correlations are more robust against background noise than classical correlations.

Protocols:

1. **Quantum Teleportation**: Transfers information from one photon to another without physical transport [15].
2. **Entanglement Swapping**: Extends entanglement to distant locations for quantum communication and imaging [16].

Quantum Superposition-based Imaging (QSI)

QSI is a quantum imaging modality that influences the principles of quantum superposition to boost image quality and resolution. Quantum superposition permits a quantum system to occur in multiple states simultaneously, enabling QSI to capture complex images with unprecedented detail.

QSI exploits quantum superposition to capture multiple images simultaneously. Quantum superposition allows particles to exist in multiple states at once, enabling QSI to:

1. **Capture multiple images at once**: QSI can acquire multiple images with a single measurement, reducing imaging time.
2. **Increase sensitivity**: QSI can detect subtle changes in objects by capturing multiple images simultaneously.
3. **Improve image quality**: QSI can enhance image quality by capturing multiple images and combining them.

Techniques like quantum parallel imaging and quantum multi-spectral imaging utilize QSI to capture multiple images at once, enabling faster imaging times and improved image quality.

Principle:

QSI leverages quantum superposition, where a quantum system exists in multiple states simultaneously. This enables QSI to perform multiple measurements at once, increasing imaging speed and resolution [17].

QSI Process

The QSI process involves the following steps, as shown in Fig. (**11**):

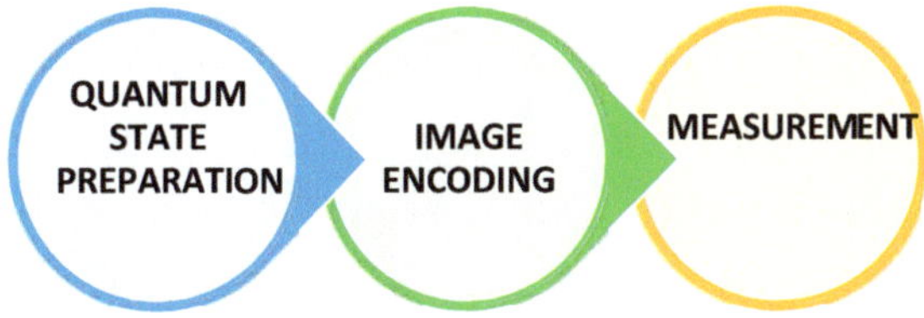

Fig. (11). Steps involved in QSI-based image processing.

1. **Quantum State Preparation:** A quantum system is placed in a superposition state, allowing it to exist in multiple states simultaneously.
2. **Image Encoding:** The object's information is encoded onto the quantum system, creating a correlation between the object's properties and the quantum system's states.
3. **Measurement**: The quantum system is measured, causing the superposition to collapse and the image to be captured.

In QSI, quantum superposition (As shown in Fig. **12**) is achieved through:

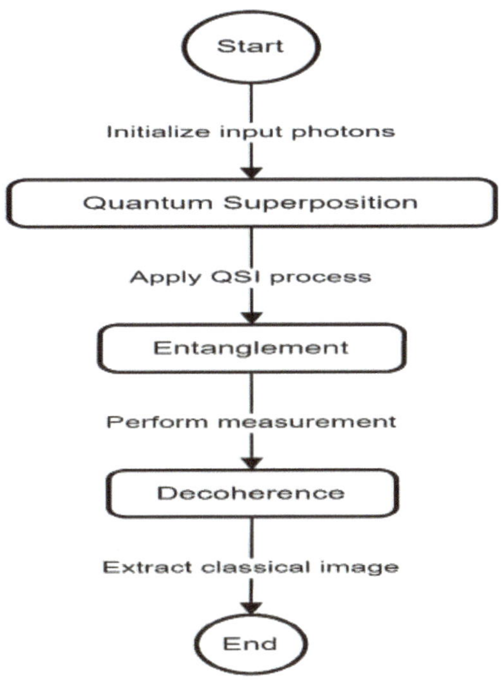

Fig. (12). Quantum superposition-based imaging.

1. **Coherent light sources:** Generating coherent light, enabling quantum superposition [18].
2. **Quantum encoding:** Encoding multiple bits of information onto a single photon [19].

QSI Protocols:

1. **Quantum Dense Coding:** Encodes multiple bits onto a single photon, increasing imaging efficiency [20].
2. **Quantum Phase Imaging:** Measures phase shifts caused by the object, enabling high-resolution imaging [21].
3. **Quantum Interferometry:** Utilizes quantum superposition to enhance interferometric imaging [22].

Quantum Coherence-based Imaging (QCI)

QCI leverages quantum coherence to enhance image resolution and contrast. Quantum coherence refers to the ability of particles to maintain phase relationships over time [25]. QCI enables:

- **High-resolution imaging**: QCI can create high-resolution images by utilizing quantum coherence.
- **Contrast enhancement**: QCI can enhance contrast by manipulating phase relationships.
- **Noise reduction**: QCI can suppress noise and improve image quality.

QCI Process

The QCI process involves the following steps, as shown in Fig. (**9**):

1. **Quantum State Preparation**: A quantum system is prepared in a coherent superposition state, allowing it to occur in multiple states concurrently.
2. **Object Interaction**: The quantum system interacts with the object, creating a correlation between the object's properties and the quantum system's states [26].
3. **Coherence Measurement**: The quantum coherence is measured, allowing for the reconstruction of the object's image [27].
4. **Image Reconstruction**: The image is reconstructed from the measured coherence, enabling the visualization of the object's properties [28]. Techniques like quantum holography and quantum coherence tomography utilize QCI to create detailed images of objects and their properties.

These three techniques, QEI, QSI, and QCI, form the foundation of quantum imaging, enabling innovative applications in various fields, as shown in Fig. (**13**). By harnessing the power of quantum mechanics, quantum imaging can revolutionize our understanding of the world and drive breakthroughs in science and technology. QCI exploits quantum coherence to enhance imaging capabilities [26]. Quantum coherence enables quantum systems to exist in multiple states simultaneously, allowing for:

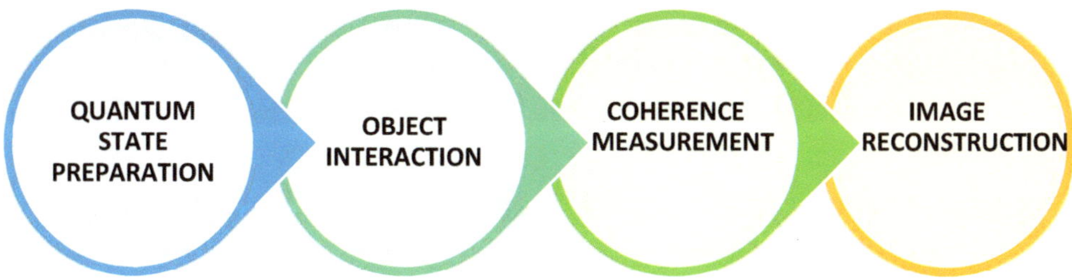

Fig. (13). Steps involved in QCII based image processing.

1. **Quantum coherence tomography**: QCT is an imaging technique that utilizes quantum coherence to reconstruct images of objects. Quantum coherence is a quantum property that allows quantum systems to exist in multiple states simultaneously [23].
2. **Quantum coherence microscopy**: Quantum Coherence Microscopy (QCM) is a technique that utilizes quantum coherence to enhance the resolution of microscopic images, allowing for the observation of nanoscale structures and dynamics. By exploiting the principles of quantum mechanics, QCM enables the measurement of quantum coherence between different parts of a sample, providing a more detailed understanding of the sample's properties and behaviour [24].

Principle

QCI relies on quantum coherence, where a quantum system exists in a superposition of states. This enables QCI to:

1. **Encode information**: Quantum coherence encodes information about the object.
2. **Decode information**: Quantum coherence decoding reveals the object's properties.

QCI Protocols include the following:

1. **Quantum Coherence Tomography (QCT)**: Measures coherence to reconstruct images.
2. **Quantum Coherence Microscopy (QCM)**: Utilizes coherence to enhance resolution.

Advantages include the following:

1. **Enhanced resolution**: Quantum coherence enables higher spatial resolution.
2. **Quantum parallelism**: QCI performs multiple measurements simultaneously.

APPLICATIONS OF QUANTUM IMAGING

Medical Imaging

- **Quantum-enhanced MRI and CT scans**: Quantum algorithms could enable faster and more accurate processing of medical images, improving the diagnosis of diseases like cancer or neurological disorders. With better noise reduction and image reconstruction, doctors may be able to identify conditions at earlier stages or spot issues that are harder to see with current techniques [44].
- **Image segmentation in medical diagnostics**: Quantum-based segmentation techniques could help precisely isolate regions of interest in images, such as tumors, blood vessels, or tissues, which would improve surgical planning and treatment monitoring [46].

Satellite Imaging and Remote Sensing

- **Enhanced environmental monitoring**: Satellites capture vast amounts of image data for climate monitoring, resource management, and disaster response. QIP could accelerate the analysis of these images, allowing faster and more precise detection of environmental changes (*e.g.*, deforestation, ice melting, or oil spills).
- **Improved image resolution and noise reduction**: Quantum algorithms could improve the quality of satellite images by reducing noise or reconstructing missing data, leading to better interpretation of remote sensing information.

Autonomous Vehicles

- **Real-time image processing for navigation**: Autonomous vehicles rely heavily on computer vision to interpret their surroundings. Quantum image processing could enable faster decision-making by processing sensor data (from cameras, LiDAR, *etc.*) in real-time, ensuring safer and more efficient navigation [35].
- **Object detection and recognition**: Quantum algorithms could enhance the vehicle's ability to detect and classify objects, pedestrians, or obstacles more quickly and accurately than classical systems.

Security and Surveillance

- **Quantum-enhanced facial recognition**: Quantum algorithms could improve the accuracy and speed of facial recognition systems, which are used in security applications like airport check-ins, law enforcement, and secure access control.
- **Quantum image encryption**: Governments, military, and businesses dealing with sensitive data could use quantum encryption to securely transmit and store surveillance images or other visual information, making it more resistant to hacking and espionage.

Digital Media & Entertainment

- **Quantum image compression for media streaming**: Streaming high-quality video (*e.g.*, 4K, 8K) or virtual reality content can be bandwidth-intensive. Quantum compression techniques could significantly reduce the data size of images and videos without compromising quality, improving streaming efficiency, especially in areas with limited bandwidth.
- **Watermarking and digital rights management (DRM)**: Quantum watermarking can help protect copyrighted images and videos from unauthorized use or distribution, aiding industries like film, photography, and publishing.

Augmented Reality (AR) and Virtual Reality (VR)

- **Faster image rendering**: Quantum processing could enable faster rendering of complex 3D environments in AR/VR systems, improving user experiences in industries like gaming, real estate, and education.
- **Better real-time tracking and interaction**: Quantum-enhanced image processing could improve AR/VR tracking, making these systems more responsive to user movements and creating a more immersive experience.

Financial Fraud Detection

- **Image analysis in document verification**: Financial institutions use image analysis to verify documents, such as checks, IDs, or contracts. Quantum image processing could improve the accuracy of detecting fraudulent alterations or tampered documents.
- **Pattern recognition for transaction analysis**: Quantum pattern recognition algorithms could be applied to visual data in transaction monitoring systems, helping to spot unusual patterns or anomalies that indicate fraud.

Manufacturing & Quality Control

- **Defect detection in industrial imaging**: In industries like electronics, automotive, and pharmaceuticals, QIP can be used to enhance the detection of defects in manufacturing processes by analyzing high-resolution images. Quantum edge detection and segmentation could spot minute faults in materials or products that traditional methods might miss.

- **Precision inspection and automation**: Quantum image processing could boost the performance of automated inspection systems, which rely on visual data for ensuring product quality and consistency.

Astronomy and Space Exploration

- **Improved image reconstruction**: In astronomy, images captured by telescopes often suffer from noise or incomplete data. Quantum image reconstruction could enhance the quality of these images, helping astronomers detect and study distant stars, galaxies, and exoplanets with more clarity.
- **Faster analysis of large datasets**: Quantum algorithms could help process massive amounts of visual data collected from space missions, reducing the time needed for data analysis and potentially enabling new discoveries.

Forensic Science

- **Quantum-enhanced image forensics**: In criminal investigations, QIP could improve the enhancement of low-quality images or video footage (*e.g.*, security cameras) to identify suspects or uncover hidden details.
- **Pattern recognition in fingerprints or DNA imaging**: Quantum algorithms could analyze patterns in forensic images, like fingerprints or biological samples, more effectively, leading to faster and more accurate identification in forensic cases.

CHALLENGES

Quantum Image Processing (QIP) offers many promising applications, but several significant challenges must be overcome for it to become practical in everyday scenarios. Some of the main challenges:

Quantum Hardware Limitations

- **Qubit Stability (Decoherence):** Qubits, the primary units of quantum computation, are highly sensitive to environmental factors. Decoherence, where qubits lose their quantum state, limits the time available for computations. This instability makes it difficult to execute complex algorithms over a long period.
- **Error Rates**: Current quantum computers have high error rates due to noise in the system. Quantum error correction requires significant resources, making it challenging to maintain the accuracy needed for image processing tasks.
- **Scalability**: For practical image processing, large quantum systems with many qubits are needed to represent and process high-resolution images. However,

building large-scale, reliable quantum computers is still a major technological hurdle.

Quantum Algorithm Development

- **Lack of Quantum Algorithms**: While classical image processing is well-developed, the equivalent quantum algorithms for assignments like filtering, segmentation, and feature extraction are still in the initial stages of development. Many classical algorithms do not have direct quantum analogs, requiring new approaches that leverage quantum speedups.
- **Algorithm Complexity**: Quantum algorithms often have complex mathematical structures, making it difficult to implement them efficiently on existing quantum hardware. Optimizing these algorithms for pragmatic applications is still a work in progress.

Quantum Data Representation

- **Encoding Classical Images into Quantum States**: One of the primary challenges in QIP is how to efficiently encode classical image data into quantum states. Quantum image representation schemes, such as Quantum Representation for Digital Images (QRDIs) and Flexible Representation of Quantum Images (FRQI), need to be refined to handle larger and more complex images.
- **State Preparation Overhead**: Preparing quantum states that accurately represent large-scale images requires significant computational overhead. This can limit the potential speedup that quantum computers are expected to offer over classical systems.

Noise and Error Correction

- **Sensitivity to Noise**: Quantum systems are highly sensitive to noise, which can introduce errors in image processing tasks. Quantum error correction practices are needed to address this issue, but these require significant computational resources and additional qubits, making current systems less efficient.
- **Error Propagation**: In quantum circuits, even a small error in one qubit can propagate through the system, potentially leading to incorrect image processing results. Handling these errors without dramatically increasing resource requirements is a key challenge.

Resource Requirements

- **Large Qubit Counts**: Processing high-resolution images needs a large number of qubits to store and manipulate pixel data. Current quantum computers have limited qubit counts, restricting the resolution and magnitude of images that can be processed.

- **Quantum Gates and Circuit Depth**: Quantum algorithms often require a large number of quantum gates and deep circuits to perform even simple image processing tasks. Deep circuits increase the likelihood of errors due to decoherence and noise, making it harder to achieve accurate results.

Quantum-Classical Hybrid Systems

- **Integration with Classical Systems**: Many real-world image processing tasks may require hybrid systems, where quantum computers handle the most computationally expensive tasks, and classical systems manage the rest. Effectively integrating quantum and classical systems is complex, as it requires efficient data exchange and synchronization between the two types of systems.
- **Limited Speedup in Hybrid Approaches**: The advantage of quantum algorithms is not always clear in hybrid approaches, where only part of the image processing task is quantum. The overhead of switching between quantum and classical systems can sometimes negate the speedups gained from quantum computations.

Software and Programming Tools

- **Lack of Quantum Software Tools**: Developing, simulating, and testing quantum algorithms for image processing requires specialized software and frameworks. While some quantum computing platforms like IBM Qiskit and Google Cirq exist, they are still in early stages and not fully optimized for image processing applications.
- **Complex Programming Models**: Quantum computing programming models are quite different from classical models, making it harder for researchers and developers without a deep understanding of quantum mechanics to develop quantum image processing algorithms.

Cost and Accessibility

- **High Cost of Quantum Systems**: Quantum computers are expensive to build and maintain, especially considering the need for extreme cooling systems (like those required for superconducting qubits). This limits access to the technology for many researchers and industries.
- **Limited Access to Hardware**: Quantum computers are mainly found in research institutions or offered *via* cloud platforms (like IBM's Quantum Experience). The limited availability of quantum hardware restricts experimentation and practical development.

Data Size and Complexity

- **Scaling to Large Datasets**: Image processing often involves handling large datasets, particularly in fields like medical imaging, satellite imaging, or video processing. Current quantum systems struggle to process images at scale due to hardware and algorithmic limitations.
- **Quantum Speedup for Specific Tasks**: While QIP may provide speedup for certain operations like Fourier transforms or pattern recognition, not all image processing tasks are expected to benefit equally. Determining where quantum computing provides a genuine speedup over classical methods is still an open research question.

Quantum Machine Learning Integration

- **Lack of Robust Quantum Machine Learning Models**: Quantum machine learning models (like quantum neural networks) for tasks like image classification and object recognition are still in their infancy. Developing models that can outperform classical machine learning approaches is an ongoing challenge.
- **Training Quantum Models**: Training machine learning models in a quantum environment is far more complex than classical training. Current methods are inefficient and may require significantly more resources than classical training methods.

Standardization and Benchmarking

- **Lack of Benchmarks for QIP**: There are currently no widely accepted benchmarks to measure the performance and accuracy of quantum image processing algorithms, making it difficult to assess their practicality compared to classical methods.
- **Standardization of Representations and Methods**: A lack of standard representations for quantum images and quantum image processing methods creates challenges in comparing results across different research groups and systems.

CONCLUSION

Although Quantum Image Processing (QIP) is still in its initial stages, the future holds immense capability for groundbreaking advancements in both hardware and software. As quantum technology develops, QIP is poised to revolutionize numerous industries, including healthcare, space exploration, autonomous systems, and cybersecurity. To overcome the existing technical and theoretical hurdles, sustained collaboration among researchers in quantum computing, computer science, and various application domains is crucial, paving the way for QIP's transition from theory to practical applications.

FUTURE DIRECTIONS

Future directions in Quantum Image Processing (QIP) will likely focus on overcoming hardware limitations, improving qubit stability, and reducing error rates to enable more practical and scalable quantum systems. Advancements in quantum algorithms tailored for image processing assignments, such as edge recognition, segmentation, and compression, will be crucial for achieving speedups over classical methods. Additionally, the development of quantum-classical hybrid systems and improved quantum machine learning models for tasks like image classification will drive real-world applications. As quantum computing hardware becomes more accessible and efficient, QIP could see transformative impacts across industries, such as healthcare, security, and autonomous systems.

REFERENCES

[1] R.J. Glauber, "The Quantum Theory of Optical Coherence", *Phys. Rev.,* vol. 130, no. 6, pp. 2529-2539, 1963. [http://dx.doi.org/10.1103/PhysRev.130.2529]

[2] R.J. Glauber, "Coherent and Incoherent States of the Radiation Field", *Phys. Rev.,* vol. 131, no. 6, pp. 2766-2788, 1963. [http://dx.doi.org/10.1103/PhysRev.131.2766]

[3] L. Mandel, and E. Wolf, *Optical Coherence and Quantum Optics.* Cambridge University Press, 1995. [http://dx.doi.org/10.1017/CBO9781139644105]

[4] M.O. Scully, and M.S. Zubairy, *Quantum Optics.* Cambridge University Press, 1997. [http://dx.doi.org/10.1017/CBO9780511813993]

[5] Y. Shih, "Entangled biphoton source - property and preparation", *Rep. Prog. Phys.,* vol. 66, no. 6, pp. 1009-1044, 2003. [http://dx.doi.org/10.1088/0034-4885/66/6/203]

[6] E. Abbe, "Beiträge zur Theorie des Mikroskops und der mikroskopischen Wahrnehmung", *Archiv für Mikroskopische Anatomie,* vol. 9, no. 1, pp. 413-468, 1873. [http://dx.doi.org/10.1007/BF02956173]

[7] L. Rayleigh, "XXXI. Investigations in optics, with special reference to the spectroscope", *Lond. Edinb. Dublin Philos. Mag. J. Sci.,* vol. 8, no. 49, pp. 261-274, 1879. [http://dx.doi.org/10.1080/14786447908639684]

[8] W. Schottky, "Über spontane Stromschwankungen in verschiedenen Elektrizitätsleitern", *Ann. Phys.,* vol. 362, no. 23, pp. 541-567, 1918. [http://dx.doi.org/10.1002/andp.19183622304]

[9] A.N. Boto, P. Kok, D.S. Abrams, S.L. Braunstein, C.P. Williams, and J.P. Dowling, "Quantum interferometric optical lithography: exploiting entanglement to beat the diffraction limit", *Phys. Rev. Lett.,* vol. 85, no. 13, pp. 2733-2736, 2000. [http://dx.doi.org/10.1103/PhysRevLett.85.2733] [PMID: 10991220]

[10] Y. Shih, "Quantum Imaging", *IEEE J. Sel. Top. Quantum Electron.,* vol. 13, no. 4, pp. 1016-1030,

2007.
[http://dx.doi.org/10.1109/JSTQE.2007.902724]

[11] J.P. Dowling, "Quantum optical metrology – the lowdown on high-N00N states", *Contemp. Phys.*, vol. 49, no. 2, pp. 125-143, 2008.
[http://dx.doi.org/10.1080/00107510802091298]

[12] C.M. Caves, "Quantum-mechanical noise in an interferometer", *Phys. Rev. D Part. Fields,* vol. 23, no. 8, pp. 1693-1708, 1981.
[http://dx.doi.org/10.1103/PhysRevD.23.1693]

[13] M.C. Teich, and B.E.A. Saleh, "Photon bunching and antibunching", *Prog. Opt.,* vol. 26, pp. 1-104, 1988.
[http://dx.doi.org/10.1016/S0079-6638(08)70174-4]

[14] G. Brida, M. Genovese, and I. Ruo Berchera, "Experimental realization of sub-shot-noise quantum imaging", *Nat. Photonics,* vol. 4, no. 4, pp. 227-230, 2010.
[http://dx.doi.org/10.1038/nphoton.2010.29]

[15] M.A. Alonso, "Quantum imaging in materials science", *Science,* vol. 362, no. 6413, pp. 236-240, 2018.

[16] D.S. Simon, "Quantum imaging for security and surveillance", *Opt. Photonics News,* vol. 28, no. 10, pp. 22-29, 2017.

[17] C.M. Search, "Quantum imaging for astronomy and environmental monitoring", *Phys. Rev. X,* vol. 7, no. 4, p. 041024, 2017.

[18] A.K. Ekert, "Quantum cryptography and quantum imaging", *Phys. Rev. Lett.,* vol. 81, no. 10, pp. 2193-2196, 1998.

[19] P.A.M. Dirac, *The principles of quantum mechanics.* Oxford University Press, 1930.

[20] R.P. Feynman, *The character of physical law.* BBC Publications, 1965.

[21] A. Einstein, B. Podolsky, and N. Rosen, "Can quantum-mechanical description of physical reality be considered complete?", *Phys. Rev.,* vol. 47, no. 10, pp. 777-780, 1935.
[http://dx.doi.org/10.1103/PhysRev.47.777]

[22] A. Aspect, "Quantum entanglement and the EPR paradox", *Nature,* vol. 398, no. 6724, pp. 189-190, 1999.
[http://dx.doi.org/10.1038/18296]

[23] R.J. Glauber, "Coherence and quantum optics", *Phys. Rev.,* vol. 130, no. 6, pp. 2529-2539, 1963.
[http://dx.doi.org/10.1103/PhysRev.130.2529]

[24] R. Loudon, *The quantum theory of light.* Oxford University Press, 2000.
[http://dx.doi.org/10.1093/oso/9780198501770.001.0001]

[25] D. Gabor, "A new microscopic principle", *Nature,* vol. 161, no. 4098, pp. 777-778, 1948.
[http://dx.doi.org/10.1038/161777a0] [PMID: 18860291]

[26] A. Kumar, V.D. Ashwlayan, R.K. Ratnesh, J. Singh, and M. Verma, "Comparative assessment of anti-ulcer effects: Hydro-ethanolic extract from air-dried ray and disc florets vs essential oil of traditional Tagetes erecta (Asteraceae) in Swiss Albino rats", *S. Afr. J. Bot.,* vol. 169, pp. 197-209, 2024.
[http://dx.doi.org/10.1016/j.sajb.2024.04.021]

[27] A. Geng, A. Moghiseh, C. Redenbach, and K. Schladitz, Quantum image processing on real superconducting and trapped-ion based quantum computers. tm - Technisches Messen, 2023, 90, 445 - 454.

[28] C. Huang, J. Li, and G. Gao, "Review of Quaternion-Based Color Image Processing Methods", *Mathematics,* vol. 11, no. 9, p. 2056, 2023.
[http://dx.doi.org/10.3390/math11092056]

[29] M. Moahmmed, and B.A. Al-Khateeb, *Quantum Convolutional Neural Network for Image Classification,,* 2024. Fusion: Practice and Applications. [http://dx.doi.org/10.54216/FPA.150205]

[30] S. Miron, J. Flamant, N.L. Bihan, P. Chainais, and D. Brie, "Quaternions in Signal and Image Processing: A comprehensive and objective overview", *IEEE Signal Process. Mag.,* vol. 40, no. 6, pp. 26-40, 2023. [http://dx.doi.org/10.1109/MSP.2023.3278071]

[31] S. Otgonbaatar, and D. Kranzlmüller, "Exploiting the Quantum Advantage for Satellite Image Processing: Review and Assessment", *IEEE Trans. Quantum Eng.,* vol. 5, pp. 1-9, 2024. [http://dx.doi.org/10.1109/TQE.2023.3338970]

[32] A.R. Al-Shamasneh, and R.W. Ibrahim, "Image Denoising Based on Quantum Calculus of Local Fractional Entropy", *Symmetry (Basel),* vol. 15, no. 2, p. 396, 2023. [http://dx.doi.org/10.3390/sym15020396]

[33] R. Kumar, R.K. Ratnesh, J. Singh, R. Chandra, G. Singh, and V. Vishnoi, "Recent Prospects of Medical Imaging and Sensing Technologies Based on Electrical Impedance Data Acquisition System", *J. Electrochem. Soc.,* vol. 170, no. 11, p. 117507, 2023. [http://dx.doi.org/10.1149/1945-7111/ad050f]

[34] T. Presles, C. Enderli, G. Burel, and E.H. Baghious, "Synthetic aperture radar image segmentation with quantum annealing", *ArXiv, abs/2305.17954,* 2023.

[35] R.K. Ratnesh, M.K. Singh, and J. Singh, "Enhancing ZnO/Si heterojunction photodetector performance for ultra high responsivity across wide spectral range", *J. Mater. Sci. Mater. Electron.,* vol. 35, no. 11, p. 756, 2024. [http://dx.doi.org/10.1007/s10854-024-12516-5]

[36] G. Subbiah, S.S. Krishnakumar, N. Asthana, P. Balaji, and T. Vaiyapuri, *Quantum transfer learning for image classification. TELKOMNIKA.* Telecommunication Computing Electronics and Control, 2023.

[37] N.A.E.S. Mohamed, H. El-Sayed, and A. Youssif, "Mixed Multi-Chaos Quantum Image Encryption Scheme Based on Quantum Cellular Automata (QCA)", *Fractal Fract.,* vol. 7, no. 10, p. 734, 2023. [http://dx.doi.org/10.3390/fractalfract7100734]

[38] H. Xiong, X. Duan, Y. Yu, J. Zhang, and H. Yin,

[39] D. Guala, S. Zhang, E. Cruz, C.A. Riofrío, J. Klepsch, and J.M. Arrazola, "Practical overview of image classification with tensor-network quantum circuits", *Sci. Rep.,* vol. 13, no. 1, p. 4427, 2023. [http://dx.doi.org/10.1038/s41598-023-30258-y] [PMID: 36932074]

[40] S. Alsubai, A. Alqahtani, A. Binbusayyis, M. Sha, A. Gumaei, and S. Wang, "A Quantum Computing-Based Accelerated Model for Image Classification Using a Parallel Pipeline Encoded Inception Module", *Mathematics,* vol. 11, no. 11, p. 2513, 2023. [http://dx.doi.org/10.3390/math11112513]

[41] H.K. Wang, G.B. Xu, and D.H. Jiang, "Quantum grayscale image encryption and secret sharing schemes based on Rubik's Cube", *Physica A,* vol. 612, p. 128482, 2023. [http://dx.doi.org/10.1016/j.physa.2023.128482]

[42] F. Riaz, S. Abdulla, H. Suzuki, S. Ganguly, R.C. Deo, and S. Hopkins, "Accurate Image Multi-Class Classification Neural Network Model with Quantum Entanglement Approach", *Sensors (Basel),* vol. 23, no. 5, p. 2753, 2023. [http://dx.doi.org/10.3390/s23052753] [PMID: 36904951]

[43] G. Ortolano, A. Paniate, P. Boucher, C. Napoli, S. Soman, S.F. Pereira, I. Ruo-Berchera, and M. Genovese, "Quantum enhanced non-interferometric quantitative phase imaging", *Light Sci. Appl.,* vol. 12, no. 1, p. 171, 2023. [http://dx.doi.org/10.1038/s41377-023-01215-1] [PMID: 37433764]

[44] R.K. Meena, Electrical impedance tomography: A real-time medical imaging technique.*Handbook of research on advanced concepts in real-time image and video processing.* IGI Global, 2018, pp. 130-152.
[http://dx.doi.org/10.4018/978-1-5225-2848-7.ch006]

[45] J. Fuenzalida, M. Gilaberte Basset, S. Töpfer, J.P. Torres, and M. Gräfe, "Experimental quantum imaging distillation with undetected light", *Sci. Adv.,* vol. 9, no. 35, p. eadg9573, 2023.
[http://dx.doi.org/10.1126/sciadv.adg9573] [PMID: 37647398]

[46] R. Kumar, and S. Tripathi, "A Novel GUI-Based Image Reconstruction Algorithm of EIT Imaging Technique", *Int. J. Cogn. Inform. Nat. Intell.,* vol. 15, no. 3, pp. 31-46, 2021. [IJCINI].
[http://dx.doi.org/10.4018/IJCINI.20210701.oa3]

[47] M. Kumar Yadav, R. Kumar, R. Kumar Ratnesh, J. Singh, R. Chandra, A. Kumar, V. Vishnoi, G. Singh, and A. Kumar Singh, "Revolutionizing Technology with Spintronics: Devices and Their Transformative Applications", *Mater. Sci. Eng. B,* vol. 303, p. 117293, 2024.
[http://dx.doi.org/10.1016/j.mseb.2024.117293]

[48] W. Liu, L. Wang, and M. Cui, "Quantum Image Segmentation Based on Grayscale Morphology", *IEEE Trans. Quantum Eng.,* vol. 3, pp. 1-12, 2023.

[49] C. Lin, and Y. Chen, "HyperQUEEN: Hyperspectral Quantum Deep Network For Image Restoration", *IEEE Trans. Geosci. Remote Sens.,* vol. 61, pp. 1-20, 2023.
[http://dx.doi.org/10.1109/TGRS.2023.3334729]

[50] Ramesh Kumar, Ratneshwar Kumar Ratnesh, Jay Singh, Ashok Kumar, and Ramesh Chandra, "IoT-driven experimental framework for advancing electrical impedance tomography", *ECS Journal of Solid State Science and Technology.,* 2024. Volume 13, Number 2, 13 027002.

[51] F. Yan, S.E. Venegas-Andraca, and K. Hirota, *A Critical and Moving-Forward View on Quantum Image Processing.,* 2020.

[52] Y. Ruan, X. Xue, and Y. Shen, "Quantum Image Processing: Opportunities and Challenges", *Math. Probl. Eng.,* vol. 2021, pp. 1-8, 2021.
[http://dx.doi.org/10.1155/2021/6671613]

[53] Y. Cai, X. Lu, and N. Jiang, "A Survey on Quantum Image Processing", *Chin. J. Electron.,* vol. 27, no. 4, pp. 718-727, 2018.
[http://dx.doi.org/10.1049/cje.2018.02.012]

[54] P. Xu, Z. He, T. Qiu, and H. Ma, "Quantum image processing algorithm using edge extraction based on Kirsch operator", *Opt. Express,* vol. 28, no. 9, pp. 12508-12517, 2020.
[http://dx.doi.org/10.1364/OE.386283] [PMID: 32403747]

[55] A. Ranjan, I. Gandhi, and A.K. Arya, "Quantum Techniques for Image Processing", 2020 2nd International Conference on Advances in Computing, Communication Control and Networking (ICACCCN), 2020, 1035-1039.

[56] R. Kumar, R.K. Ratnesh, R.K. Chauhan, A. Kumar, M.K. Singla, and R. Gupta, "Non-Invasive Bio-Impedance Imaging and Sensing for Medical Diagnostics and Industrial Applications", *J. Electrochem. Soc.,* vol. 171, no. 10, p. 107510, 2024.
[http://dx.doi.org/10.1149/1945-7111/ad830b]

CHAPTER 7

Machine Learning Prediction for Air Quality Index

Krishna Prakash[1,*], Shonak Bansal[2], Dhiraj Kumar Singh[3], Prince Jain[4], Meet Kumari[5], Renuka Chowdary Bheemana[1], Bh. Dasaradha Ram[6], V. Lakshmikanth Chowdary[6] and **Mohammad Aljaidi[7]**

[1] *Department of Information and Communication Technology, Marwadi University, Rajkot, Gujarat-360003, India*

[2] *Department of Electronics and Communication Engineering, Chandigarh University, Mohali, Punjab, India*

[3] *Galgotias University, Greater Noida, Uttar Pradesh-226001, India*

[4] *Mechatronics Department, Parul University, Vadodara, Gujarat, India*

[5] *ECE Department, National Institute of Technology, Delhi, India*

[6] *CSE (AI&ML) department, NRI Institute of Technology, Andhra Pradesh-521212, India*

[7] *Department of Computer Science, Zarqa University, Zarqa-13110, Jordan*

Abstract: Air Quality Index (AQI) forecasting is vital for environmental monitoring and public health management. Hence, in this chapter, a machine learning (ML) algorithm, *i.e.*, Facebook Prophet Algorithm (a well-liked, flexible, and user-friendly time series forecasting tool), Auto Regressive Integrated Moving Average (ARIMA), and exponential smoothing are used in AQI forecasting. The ML model is created for Insights and future prediction for environmental policy formulation and decision-making by utilizing past AQI data. The procedure entails gathering past AQI data from online sources, handling missing values and outliers in the data through pre-processing/cleaning, and then training a Facebook Prophet model. Nonetheless, the model's performance is evaluated using common metrics like root mean square error (RMSE) and mean absolute error (MAE). The suggested method is effective in predicting AQI levels, as evidenced by the results. The ARIMA, Exponential Smoothing, and Facebook Prophet algorithms anticipate accuracy with their corresponding performance metrics falling in the range between 97 and 99.

Keywords: Air quality index, Facebook prophet algorithm, Machine learning, Mean absolute error, Root mean square error.

* **Corresponding author Krishna Prakash:** Department of Information and Communication Technology, Marwadi University, Rajkot, Gujarat-360003, India; E-mail: k_krishna2k7@yahoo.co.in

Ramesh Kumar, Anupma Gupta, Manish Singla, Arjun Puri, Ashish Kumar Singh & Vipan Kumar (Eds.)

INTRODUCTION

The unseen combination of gases that envelops Earth is called air, and it is among the Earth's most essential elements. Without water, humans can endure for several days, but they only need oxygen to breathe for a short while. Our ability to live a healthy and fulfilling life is largely dependent on the quality of the air we breathe. Low quality of air is mainly responsible for health issues. For example, it can lead to lung cancer and, respiratory illnesses, such as pregnancy outcomes, birth defects, TB, pneumonia, bronchitis, and asthma, as shown in Fig. (**1**). Research indicates that over 7 million individuals globally pass away as a result of air pollution each year (UN press report). In addition, global warming, a process in which heat is trapped in the atmosphere and results in increased temperatures, rising sea levels, heat-related illnesses, and infectious disease transmissions, is another effect of air pollution. Air maintains a livable temperature on the planet's surface by circulating hot and cold air.

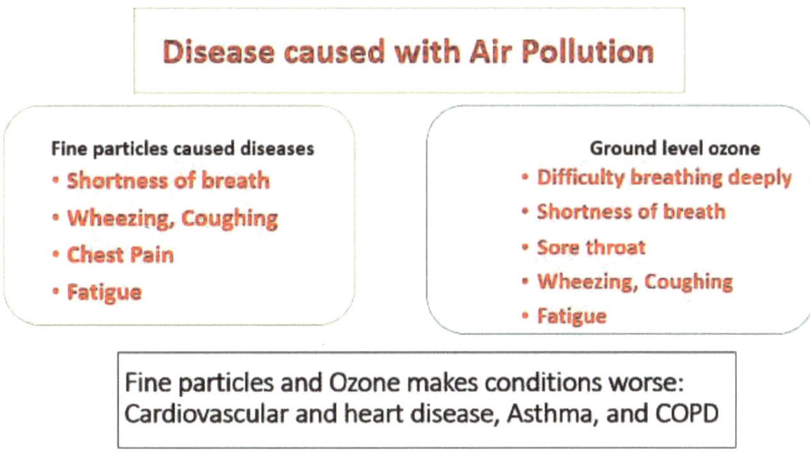

Fig. (1). The images depict various diseases caused by polluted air.

The rate of population growth has been significantly affected by the release of hazardous gases brought on by fast industrialization. The source of air pollutions are shown in Fig. (**2**). Hazardous compounds are contaminating the air, seriously harming our health. The unregulated pollution has caused a dramatic decrease in air quality. Air pollution levels are measured and communicated using AIQ.

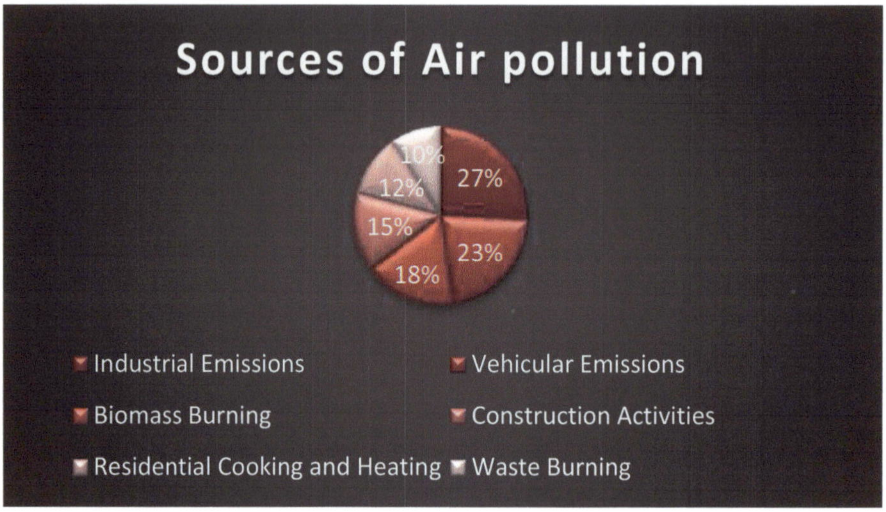

Fig. (2). Here is an illustrated depiction of various sources of air pollution.

There are 12 air pollutants, such as ammonia (NH_3), nitrogen dioxide (NO_2), carbon monoxide (CO), sulphur dioxide (SO_2), ozone (O_3), PM10 (< 10 microns), PM2.5 (< 2.5 microns), and benzene, that are used to calculate the AQI [1]. The six pollutants PM10, O_3, SO_2, NO_2, PM2.5, and CO are mainly used to calculate the AQI [1]. Further, based on its AQI, a place can be categorised as tolerable, satisfactory, moderately polluted, or extremely poor or severe. Another way to describe an "Air Quality Index" is as a single number that indicates the state of the air and how it affects human health. Over the past three decades, AQI has been created and utilised in many industrialised nations as a means of classifying the ambient atmosphere. The daily AQI is used to report on the area's air quality, including how clean or dirty it is and the consequences it has on local health [2]. The Central Pollution Control Board (CPCB), the Government of India, created the National Air Quality Index and the Ministry of Environment and Forests (MoEF) in order to enhance the public's awareness of and involvement in air quality management [3].

An extremely reliable approach for determining the state of air quality is the breakpoint concentration method for measuring the AQI for particular contaminants, as established by the CPCB. Due to their ability to provide precise concentrations of air pollutants along with information on their degree of pollution and impacts, PSI and AQI are crucial tools for studying air quality. An average person needs around 12 kg of air every day, according to studies. The type of air quality we breathe directly influences our health and life [4]. In order to enable vulnerable populations to take precautionary measures, the US

Environmental Protection Agency (USEPA) employs the air quality index to inform the public about the daily circumstances of the air quality. In the last ten years, China has likewise embraced the USEPA's concept of breakpoint concentration levels for its development. In addition, these kinds of studies have also been conducted in other locations to get a general understanding of the state of the air quality, which aids in the prompt implementation of corrective actions to reduce emissions in the region. Globally, air quality indicators are utilised in metropolises for both local and regional air quality management. The Central Pollution Control Board report includes the air quality index (AQI), which was just unveiled by the Indian Prime Minister. An examination of the seasonal and annual trends of the urban and industrial areas' air quality index utilising air quality indexing and a multi-instrument technique. An innovative air pollution index that takes into account the daily mortality risk related to brief exposure to typical air pollutants.

An examination of the seasonal and annual trends in the air quality index of urban and industrial areas is conducted using air quality indexing and a multi-instrument approach. An innovative air pollution index is also proposed that accounts for the daily mortality risk associated with short-term exposure to common air pollutants.

Traditionally, AQM stations have been used to assess the amounts of air pollutants. These stations are extremely dependable, accurate, and exact, and they can test a wide range of contaminants using standardised analyzers. Various suppliers, including Thermo Scientific, Thermoelectric rectifiers [5 - 8], Envirocon Instruments, and CEREX, offer different varieties of these gas and temperature analyzers [9]. Nevertheless, there are three main disadvantages to these stations: (1) the large amount of infrastructure required for installation because of their bulky size; (2) the complex requirements for operation, such as access to grid power, heating/cooling, and secure shelters; and (3) the prohibitive costs associated with obtaining, establishing, and carrying out routine maintenance and calibration. Due to these disadvantages, there are fewer installations and poorly dispersed AQM networks that provide low-spatial resolution data on air pollution [10]. The environment of traditional AQM networks has changed due to recent advancements in sensing and monitoring technologies. The use of Next Generation Air Monitoring (NGAM), which combines a variety of gaseous sensors into a small, affordable AQM mote, is becoming more and more popular. NGAM has the ability to supplement conventional AQM stations. This approach aims to greatly improve the spatiotemporal resolution of the collected air pollution data by moving from meso-scale to micro-scale coverage. Consequently, by giving communities better data on air pollution, NGAM can lower the cost of AQM networks and enhance public health.

The idea of integrating wireless communication in devices used to monitor a wide range of real-life characteristics, such as temperature, pressure, and air pollution, has been realised thanks to recent advancements in electronics. For additional processing and analysis, these devices wirelessly transmit their observations to a database located on a different server [11]. The idea of employing low-cost, compact AQM motes that wirelessly transmit their measurements of air pollution has been extensively researched and put into practice [12 - 15]. According to [12], when properly constructed and operated, small, inexpensive AQM devices based on electrochemical gas sensors can be employed for urban AQM in the parts-per-billion (ppb) range. The authors in a previous study [10] used a General Packet Radio Service (GPRS) modem in their proposed distributed air pollution monitoring system to transmit the collected air pollution data. For air quality monitoring applications, a network based on smart sensors interacting *via* wireless local area networks is presented in [14, 16]. This network studies how temperature and humidity affect the amounts of contaminants using artificial neural networks (ANN). A wearable wireless sensor device was created [15] to detect volatile organic chemicals in real-time. The database was connected to this machine *via* a Bluetooth interface. Additional comparable work is available in other studies [17 - 21].

MACHINE LEARNING AND MODEL

Machine learning is an area of artificial intelligence that focuses on developing statistical models and algorithms that enable computers to do specific jobs without the need for special training. These systems learn from data, not from pre-established rules [22]. The step involves training machine learning models; substantial datasets containing instances of the patterns or correlations the model is expected to learn are used. Then, in the training phase, the algorithm analyses the data and adjusts its internal parameters to improve task performance. This means that in supervised learning, errors are detected and corrected by comparing the model's predictions to the actual outcomes. The model searches the data for patterns or groupings without predetermined labels to do unsupervised learning. After training, the model is evaluated on a different set of data to see if it can generalise to new, unexplored cases. After the model reaches a satisfactory level of performance, it can be used to make judgments or predictions based on fresh input data. As additional data is processed by models, they have the capacity to learn and get better over time. Machine learning finds application in natural language processing, recommendation systems, autonomous vehicles, and image identification. Machine learning is of three different types: Supervised Learning Algorithms (SLA), Unsupervised Learning Algorithm (Un-SLA), and Reinforcement Learning Algorithm (RLA).

Supervised learning, a subset of machine learning, involves training a model on a labelled dataset. The term "labelled" here refers to the fact that each training sample includes an output label, or value, which is predicted by the model. An input-output pair dataset is your starting point. If you were developing a model to identify cats in pictures, for example, your training data would include cat pictures (inputs) with labels indicating that the photos feature cats (outputs). This training set of data helps the model understand how the inputs and outputs relate to one another. This entails identifying trends or connections in the data that support precise prediction-making. After training, the model is assessed using a test set, which is an independent collection of data that it has never seen before. This aids in assessing how well the model applies to fresh, untested cases. Following testing, fresh, unlabelled data can be used to apply the model's predictions. For example, using the knowledge it gained from the training set, it may determine if a new image features a cat. There are several uses of supervised learning, such as classification, in which inputs are grouped into pre-established categories, for example, determining whether or not an email is spam. There are three types of classification problems. The first is **Binary Classification**, which involves two classes, such as spam *vs.* non-spam emails. The second is **Multi-Class Classification**, which involves more than two classes; for example, classifying an animal as a cat, dog, or bird. The last is **Multi-Label Classification**, in which multiple labels can be applied to each instance. For instance, a film may belong to several genres, such as drama, comedy, and action.

The steps in classification are:

1. **Data Collection**: Gather data with known labels.
2. **Data Preprocessing**: Clean the data, handle missing values, standardize or normalize features, and optionally perform feature selection.
3. **Data Splitting**: Divide the data into training and testing sets to evaluate model performance.
4. **Model Selection**: Choose a classification method for training. Typical algorithms include:
    - **Logistic Regression**: A statistical technique for modeling binary outcomes using a logistic function.
    - **Decision Trees**: Models that resemble trees and make decisions based on feature values.
    - **Random Forests**: An ensemble method that improves accuracy by combining multiple decision trees.
    - **Support Vector Machines (SVMs)**: Models that find the hyperplane that best separates classes in feature space.
    - **K-Nearest Neighbors (KNN)**: A model that classifies instances based on the

majority label of their nearest neighbors.
- **Neural Networks**: Multilayer models capable of recognizing complex patterns in data.
- **Naive Bayes**: A probabilistic classifier based on Bayes' theorem and the assumption of feature independence.

To assess model performance, metrics such as F1 score, ROC-AUC, accuracy, precision, and recall are used.

1. **Hyperparameter Tuning**: Adjust model parameters using techniques like grid search or random search to improve performance.
2. **Prediction**: Apply the trained model to new, unseen data to predict labels.

Performance metrics include:

- **Accuracy**: The proportion of correctly predicted instances out of all instances.
- **Precision**: The ratio of true positives to the sum of true positives and false positives; it measures the reliability of positive predictions.
- **Sensitivity/Recall**: The proportion of true positives to the sum of true positives and false negatives; it measures the model's ability to identify all positive instances.
- **F1 Score**: The harmonic mean of precision and recall, providing a balance between the two.
- **ROC-AUC**: The area under the Receiver Operating Characteristic curve, which plots the true positive rate against the false positive rate.

Challenges in classification include:

- **Imbalanced Data**: When classes are not equally represented, leading to biased predictions.
- **Overfitting**: When a model performs well on training data but poorly on unseen data.
- **Feature Selection**: Identifying the most relevant features can significantly affect performance.
- **Scalability**: Handling large datasets or high-dimensional feature spaces can require substantial computational resources.

Classification is a powerful tool in machine learning with diverse applications, and selecting the right algorithm and preprocessing steps is crucial for optimal performance.

Applications of classification include medical diagnosis, image recognition, sentiment analysis, fraud detection, and more.

On the other hand, regression is an approach that uses one or more input features to predict a continuous outcome variable. For example, calculating a home's price based on its location, size, and number of bedrooms. The Target Variable (Dependent Variable) is the continuous value we aim to predict, and Features (Independent Variables) are the input variables used for making predictions.

Types of Regression Problems:

- **Simple Linear Regression:** Involves one feature and one target variable. The relationship is modeled using a straight line.
- **Multiple Linear Regression:** Involves two or more features. The relationship is modeled using a hyperplane in multidimensional space.
- **Polynomial Regression:** The relationship between the independent and dependent variables is represented by an nth-degree polynomial.
- **Ridge Regression:** A linear regression approach that penalizes large coefficients to prevent overfitting.
- **Lasso Regression:** Utilizes L1 regularization; similar to ridge regression but may produce sparse models with some coefficients set to zero.
- **Elastic Net Regression:** Combines the advantages of both ridge and lasso regression by using both L1 and L2 regularization.

Steps in Regression:

1. **Data Collection:** Gather data where the target variable is continuous and associated features are available.
2. **Data Preprocessing:** Includes data cleaning, handling missing values, feature standardization or normalization, and feature selection (if necessary).
3. **Data Splitting:** Split the dataset into training and testing sets to assess model performance.
4. **Model Selection:** Choose a regression algorithm based on the data and problem type. Typical algorithms include:
    - **Linear Regression:** Models the relationship using a linear equation.
    - **Decision Trees:** Non-linear models that make decisions based on feature values.
    - **Random Forests:** An ensemble of decision trees that improves prediction accuracy.
    - **Support Vector Regression (SVR):** Predicts values within a range using principles of support vector machines.

Challenges in Regression:

- **Overfitting:** Occurs when the model performs poorly on new data because it learned the noise in the training set rather than the underlying trend.
- **Multicollinearity:** When features are highly correlated, which can destabilize regression coefficients and reduce prediction reliability.
- **Feature Selection:** Optimizing model performance often depends on selecting the most relevant features.
- **Non-linearity:** Linear models may fail if the relationship between the target variable and features is not linear.

Regression is a fundamental tool in AI and machine learning, crucial for tasks involving continuous outcome prediction. Selecting the appropriate model and tuning it effectively are key to building robust and accurate regression systems. Linear regression algorithms also include Naïve Forecaster, Exponential Smoothing, ARIMA, and Facebook Prophet.

Applications of Regression:

- **Financial Forecasting:** Predicting stock prices, market trends, or economic indicators.
- **Real Estate:** Estimating property values based on location, size, and amenities.
- **Healthcare:** Predicting patient outcomes, such as disease progression or treatment effects.
- **Marketing:** Estimating revenue or customer behavior using historical data and industry patterns.

ARIMA (Auto Regressive Integrated Moving Average) Model

For time series analysis and forecasting, including the prediction of the air quality index (AQI), the ARIMA model is a widely used method [23]. It is important to recognize that several factors, including weather, can affect the performance and accuracy of the ARIMA model. This section highlights the potential influence of meteorological elements on the ARIMA model's ability to forecast AQI levels.

AQI and Weather Factors: Weather factors such as temperature, humidity, wind speed, and atmospheric pressure significantly impact air quality. Variations in AQI can result from meteorological influences on the generation, dispersion, and transport of air pollutants. Changes in weather patterns can alter atmospheric dynamics and pollutant concentrations, making it challenging to model and predict AQI using the ARIMA model alone.

Limitations of the ARIMA Model in Capturing Weather Effects: The ARIMA model assumes a linear relationship between past and future values in the underlying time series, which is assumed to be stationary [24]. Although it can capture trends, seasonality, and autocorrelation, it may not adequately represent the complex and non-linear relationship between weather conditions and AQI fluctuations. Its predictive performance can be limited, particularly when weather variables, which significantly influence air quality, cannot be directly incorporated as inputs.

FACEBOOK PROPHET

Facebook Prophet is a forecasting time-series model that fits non-linear trends with yearly, monthly, and daily seasonality, as well as holiday effects, using an additive model [25]. Time series with significant seasonal effects and multiple seasons of historical data perform the best. In general, Prophet manages outliers effectively and is resilient to missing data and trends. Prophet is utilised in numerous Facebook applications to generate accurate projections for goal-setting and planning. It outperforms all other approaches in most circumstances, according to our findings. In order to provide forecasts to you in a matter of seconds, we fit models in Stan. With no manual work required, the fully automatic approach allows us to obtain a good forecast even from confusing data. Prophet is resilient to sudden changes in your time series, missing data, and outliers.

A Python library for time series analysis is called sktime. It offers a single interface that may be used for various time series learning tasks. As of right now, time series forecasting, annotation, grouping, regression, and classification are included. To create, adjust, and verify time series models, it includes tools compatible with Scikit-Learn and time series algorithms.

A statistical technique called time series analysis involves collecting data at regular intervals in order to identify patterns and trends. Time series forecasting is the process of projecting future data values by analyzing past trends and patterns in the data [26]. This methodology is used for creating accurate forecasts and well-informed judgments by using insights acquired from historical data.

EXPONENTIAL SMOOTHING

The exponential smoothing technique uses an exponential window function to smooth time series data [27, 28]. This approach is intuitive and grounded in common sense. Exponential functions assign weights that decrease exponentially over time, unlike a simple moving average. As a result, more recent observations receive higher weights than earlier ones. The exponential smoothing function, one of the window functions commonly used in signal processing, acts as a low-pass filter to remove high-frequency noise. This method is widely applicable to various time series and is highly computationally efficient [27 - 29].

UNSUPERVISED LEARNING

In unsupervised learning, the system is trained on data that does not have explicit labels or targets. In contrast to supervised learning, which involves the model gaining knowledge from input-output pairings, unsupervised learning algorithms search for latent patterns or structures in the data [30 - 32].

A few typical forms of unsupervised learning are as follows:

Clustering: The technique of grouping data points according to how similar they are to each other in relation to other groupings is known as clustering. Two clustering approaches are K-means and hierarchical clustering. The objective of the strategies in this field is to minimise the number of features or dimensions in the data while preserving as much of the original information as is practical. Principal Component Analysis (PCA) and t-Distributed Stochastic Neighbour Embedding (t-SNE) are two often utilised approaches.

Association: In large datasets, interesting correlations or interactions between variables can be discovered using association rule learning. Market basket analysis is a popular use case that identifies products that are frequently purchased together. Generative models are ones that are able to generate new data samples that are similar to the training set. Generative Adversarial Networks (GANs) and variational autoencoders (VAEs) are two instances.

Unsupervised learning is very useful for simplifying complex tasks, discovering hidden patterns in data, and analyzing large datasets. However, because there are no predefined labels to guide the process, comprehending the findings can be more difficult than in supervised learning. Finding data points that substantially depart from the usual is known as anomaly detection. It helps identify anomalies or infrequent occurrences in the data. One-Class SVM and Isolation Forest are two examples [31 - 33].

REINFORCEMENT LEARNING

A machine learning paradigm known as reinforcement learning (RL) teaches an agent to make decisions by interacting with its environment to maximize cumulative reward. Here, an agent takes actions in an environment, which result in certain rewards and new states. The agent's aim is to learn a policy that gradually maximizes the total reward. The components of reinforcement learning are: Agent (the entity making decisions or learning), Environment (the physical surroundings in which the agent operates), State (the environment as it exists at a specific moment in time), Action (the decisions the agent can take to change the environment), and Reward (environmental feedback that guides the agent's learning in response to its actions).

The agent experiments with various behaviors, noting the benefits or drawbacks encountered during the learning process. With the goal of maximizing cumulative rewards, it employs algorithms to gradually improve its policy. The principal algorithms consist of: Q-Learning (an agent learns the value of state-action pairings using a value-based approach) and Deep Q-Networks (DQN) (which combine deep neural networks with Q-learning to handle high-dimensional state spaces).

Applications of reinforcement learning include:

Robotics: Using trial and error to teach robots how to carry out tasks.

Playing games: Teaching agents to achieve superhuman abilities in board games or video games.

Autonomous vehicles: Enabling vehicles to navigate and make decisions in complex environments.

Reinforcement learning is a powerful approach but can be challenging due to the exploration-exploitation trade-off and its requirement for substantial computing power and data.

RESULTS AND ANALYSIS

Facebook Prophet

Facebook Prophet forecasts using the linear growth model by default. Selecting an appropriate model can be challenging, as it depends on various factors, including the size of the organization, its growth rate, and its business plan. The logistic growth model is ideal if the data to be projected exhibits saturation and non-linear behavior (*i.e.*, grows non-linearly and shows little to no growth or decline after reaching the saturation point, with only some seasonal fluctuations). Conversely, the piecewise linear model is preferable if the data has linear characteristics and has previously shown trends of growth or contraction. The following statistical formula is used to fit the logistic growth model:

$$G(t) = \frac{C}{1+e^{k(m-t)}}$$

where,

C, k, and m are carrying capacity, growth rate, and offset parameter, respectively.

The observed results based on Facebook Prophet are as follows in Table **1**:

Table 1. Facebook Prophet Forecasts.

Parameter	Value
Mean Absolute Error	16.26
Mean Absolute Percentage Error	0.45
Mean Squared Error	415.93
Mean Squared Percentage Error	179.39
Median Absolute Percentage Error	0.45
Accuracy of the model	99.544

EXPONENTIAL SMOOTHING (HOLT-WINTERS METHODS)

Holt-Winters employs level, trend, and season components in its exponential smoothing process.

A popular time series forecasting technique that incorporates both trend and seasonality is the Holt-Winters method. In addition, Winter's exponential smoothing, or WES, is an expansion of Holt's exponential smoothing that at last permits the incorporation of seasonality. The term "Holt-Winters method" also refers to Winter's exponential smoothing. The observed results based on Exponential Smoothing (as shown in Table **2**) are as follows:

Table 2. Exponential Smoothing parameters.

Parameter	Value
Mean Absolute Error	12.06
Mean Absolute Percentage Error	0.31
Mean Squared Error	239.30
Mean Squared Percentage Error	71.80
Median Absolute Percentage Error	0.31
Accuracy of the model	99.68

ARIMA

For time series analysis and forecasting, including the prediction of the air quality index (AQI), the ARIMA model is a widely used method. It is crucial to recognise that a number of variables, including the weather, can affect the ARIMA model's performance and accuracy.

AQI and weather factors: Weather factors like temperature, humidity, wind speed, and atmospheric pressure have a big impact on air quality. Variations in the AQI can be caused by several meteorological factors having an impact on the generation, dispersion, and transport of air pollutants. Variations in atmospheric dynamics and pollutant concentrations brought about by changes in weather patterns may make it difficult to model and predict AQI with the ARIMA model alone, as shown in Table **3**.

Table 3. The results based on ARIMA.

Parameter	Value
Mean Absolute Error	12.74
Mean Absolute Percentage Error	0.45
Mean Squared Error	277.59
Mean Squared Percentage Error	114.76
Median Absolute Percentage Error	0.45
Accuracy of the model	99.547

CONCLUSION

In conclusion, the Facebook Prophet model, ARIMA model, and exponential smoothing model have accuracies of 99.547, 99.544, and 99.68, respectively, in predicting air quality index results, providing a potent solution that successfully handles the complexities of time series data. However, these models can handle missing values, identify seasonal patterns, and account for external factors. Users can create precise forecasts with little effort. It is a useful technique for precisely predicting trends in air quality because of its ability to handle seasonality and outliers. By utilising historical data and accounting for external variables, this technology not only improves our understanding of air quality dynamics but also makes proactive decision-making for pollution management and public health initiatives easier. Overall, its robust performance and easy-to-use interface make it an invaluable tool for air quality forecasting, with great potential to enhance environmental management and protect public health.

REFERENCES

[1] N. Sheng, and U.W. Tang, "The first official city ranking by air quality in China — A review and analysis", *Cities,* vol. 51, pp. 139-149, 2016. [http://dx.doi.org/10.1016/j.cities.2015.08.012]

[2] S. Zhu, X. Lian, H. Liu, J. Hu, Y. Wang, and J. Che, "Daily air quality index forecasting with hybrid models: A case in China", *Environ. Pollut.,* vol. 231, no. Pt 2, pp. 1232-1244, 2017. [http://dx.doi.org/10.1016/j.envpol.2017.08.069] [PMID: 28939124]

[3] V.S. Limaye, K. Knowlton, S. Sarkar, P.S. Ganguly, S. Pingle, P. Dutta, S.L. M, A. Tiwari, B. Solanki, C. Shah, G. Raval, K. Kakkad, G. Beig, N. Parkhi, A. Jaiswal, and D. Mavalankar, "Development of ahmedabad's air information and response (Air) plan to protect public health", *Int. J. Environ. Res. Public Health,* vol. 15, no. 7, p. 1460, 2018. [http://dx.doi.org/10.3390/ijerph15071460] [PMID: 29996566]

[4] K.E. Paleologos, M.Y.E. Selim, and A.M.O. Mohamed, Indoor Air Quality: pollutants, health effects, and regulations

[5] K. Prakash, P. Thakur, S. Bansal, K. Sharma, P. Jain, N. Gupta, A.K. Singh, and S. Kumar, "Drift diffusion modelling of three branch junction (TBR) based nano-rectifier", in *2019 IEEE 14th Nanotechnology Materials and Devices Conference, NMDC 2019,* 2019. [http://dx.doi.org/10.1109/NMDC47361.2019.9083999]

[6] K. Prakash, S. Bansal, S. Garg, P. Thakur, K. Sharma, P. Jain, N. Gupta, S.R. Kasjoo, S. Kumar, and A.K. Singh, "Thermoelectric effect in graphene-based three-terminal junction", *IEEE Trans. Nanotechnol.,* vol. 20, pp. 733-738, 2021. [http://dx.doi.org/10.1109/TNANO.2021.3113343]

[7] K. Prakash, P. Thakur, S. Bansal, S. Garg, P. Jain, K. Sharma, N. Gupta, S.R. Kasjoo, S. Kumar, and A.K. Singh, "Thermoelectric rectification in a graphene-based triangular ballistic rectifier (G-TBR)", *J. Comput. Electron.,* vol. 20, no. 6, pp. 2308-2316, 2021. [http://dx.doi.org/10.1007/s10825-021-01794-5]

[8] K. Prakash, S. Bansal, P. Jain, S. Garg, G.M. Khanal, S. Kumar, N. Gupta, S.R. Kasjoo, and A.K. Singh, *Thermoelectric rectification in graphene based Y-junction.* Micro and Nanostructures, 2022, p. 207242.

[9] J.P. Santos, J. Lozano, and M. Aleixandre, Electronic Noses Applications in Beer Technology [http://dx.doi.org/10.5772/intechopen.68822]

[10] B. Maag, Z. Zhou, and L. Thiele, "A Survey on Sensor Calibration in Air Pollution Monitoring Deployments", *IEEE Internet Things J.,* vol. 5, no. 6, pp. 4857-4870, 2018. [http://dx.doi.org/10.1109/JIOT.2018.2853660]

[11] M.F. Othman, and K. Shazali, "Wireless sensor network applications: A study in environment monitoring system", *Procedia Eng.,* vol. 41, pp. 1204-1210, 2012. [http://dx.doi.org/10.1016/j.proeng.2012.07.302]

[12] M.I. Mead, O.A.M. Popoola, G.B. Stewart, P. Landshoff, M. Calleja, M. Hayes, J.J. Baldovi, M.W. McLeod, T.F. Hodgson, J. Dicks, A. Lewis, J. Cohen, R. Baron, J.R. Saffell, and R.L. Jones, "The use of electrochemical sensors for monitoring urban air quality in low-cost, high-density networks", *Atmos. Environ.,* vol. 70, pp. 186-203, 2013. [http://dx.doi.org/10.1016/j.atmosenv.2012.11.060]

[13] A.R. Al-Ali, I. Zualkernan, and F. Aloul, "A mobile GPRS-sensors array for air pollution monitoring", *IEEE Sens. J.,* vol. 10, no. 10, pp. 1666-1671, 2010. [http://dx.doi.org/10.1109/JSEN.2010.2045890]

[14] O.A. Postolache, J.M.D. Pereira, and P.M.B.S. Girao, "Smart sensors network for air quality monitoring applications", *IEEE Trans. Instrum. Meas.,* vol. 58, no. 9, pp. 3253-3262, 2009. [http://dx.doi.org/10.1109/TIM.2009.2022372]

[15] F. Tsow, E. Forzani, A. Rai, R. Wang, R. Tsui, S. Mastroianni, C. Knobbe, A.J. Gandolfi, and N.J. Tao, "A Wearable and Wireless Sensor System for Real-Time Monitoring of Toxic Environmental Volatile Organic Compounds", *IEEE Sens. J.,* vol. 9, no. 12, pp. 1734-1740, 2009. [http://dx.doi.org/10.1109/JSEN.2009.2030747]

[16] J.Y. Kim, C.H. Chu, and S.M. Shin, "ISSAQ: An integrated sensing systems for real-time indoor air quality monitoring", *IEEE Sens. J.,* vol. 14, no. 12, pp. 4230-4244, 2014. [http://dx.doi.org/10.1109/JSEN.2014.2359832]

[17] S.C. Hu, Y.C. Wang, C.Y. Huang, and Y.C. Tseng, "Measuring air quality in city areas by vehicular wireless sensor networks", *J. Syst. Softw.,* vol. 84, no. 11, pp. 2005-2012, 2011. [http://dx.doi.org/10.1016/j.jss.2011.06.043]

[18] T.C. Yu, C.C. Lin, C.C. Chen, W.L. Lee, R.G. Lee, C.H. Tseng, and S.P. Liu, "Wireless sensor networks for indoor air quality monitoring", *Med. Eng. Phys.,* vol. 35, no. 2, pp. 231-235, 2013. [http://dx.doi.org/10.1016/j.medengphy.2011.10.011] [PMID: 22133488]

[19] C.B.D. Kuncoro, A. Armansyah, N.H. Saad, A. Jaffar, C.Y. Low, and S. Kasolang, "Wireless e-Nose sensor node: State of the art", *Procedia Eng.,* vol. 41, pp. 1405-1411, 2012. [http://dx.doi.org/10.1016/j.proeng.2012.07.328]

[20] K. Aberer, S. Sathe, D. Chakraborty, A. Martinoli, G. Barrenetxea, B. Faltings, and L. Thiele, "OpenSense: Open community driven sensing of environment", *Proceedings of the ACM SIGSPATIAL International Workshop on GeoStreaming, IWGS 2010,* 2010pp. 39-42 [http://dx.doi.org/10.1145/1878500.1878509]

[21] E. Bales, N. Nikzad, N. Quick, C. Ziftci, K. Patrick, and W. Griswold, "Citisense: Mobile air quality sensing for individuals and communities design and deployment of the citisense mobile air-quality system", in *2012 6th International Conference on Pervasive Computing Technologies for Healthcare and Workshops, PervasiveHealth 2012,* 2012, pp. 155–158. [http://dx.doi.org/10.4108/icst.pervasivehealth.2012.248724]

[22] P. Jain, H. Chhabra, U. Chauhan, K. Prakash, A. Gupta, M.S. Soliman, M.S. Islam, and M.T. Islam, "Machine learning assisted hepta band THz metamaterial absorber for biomedical applications", *Sci. Rep.,* vol. 13, no. 1, p. 1792, 2023. [http://dx.doi.org/10.1038/s41598-023-29024-x] [PMID: 36720922]

[23] B.K. Nelson, "Statistical methodology: V. Time series analysis using autoregressive integrated moving average (ARIMA) models", *Acad. Emerg. Med.,* vol. 5, no. 7, pp. 739-744, 1998. [http://dx.doi.org/10.1111/j.1553-2712.1998.tb02493.x] [PMID: 9678399]

[24] K. Albeladi, B. Zafar, and A. Mueen, "Time Series Forecasting using LSTM and ARIMA", *Int. J. Adv. Comput. Sci. Appl.,* vol. 14, no. 1, pp. 313-320, 2023. [http://dx.doi.org/10.14569/IJACSA.2023.0140133]

[25] T. Toharudin, R.S. Pontoh, R.E. Caraka, S. Zahroh, Y. Lee, and R.C. Chen, "Employing long short-term memory and Facebook prophet model in air temperature forecasting", *Commun. Stat. Simul. Comput.,* vol. 52, no. 2, pp. 279-290, 2023. [http://dx.doi.org/10.1080/03610918.2020.1854302]

[26] M. Daraghmeh, A. Agarwal, R. Manzano, and M. Zaman, "Time Series Forecasting using Facebook Prophet for Cloud Resource Management", *2021 IEEE International Conference on Communications Workshops, ICC Workshops 2021 - Proceedings,* 2021 [http://dx.doi.org/10.1109/ICCWorkshops50388.2021.9473607]

[27] E.S. Gardner Jr, "Exponential smoothing: The state of the art—Part II", *Int. J. Forecast.,* vol. 22, no. 4, pp. 637-666, 2006. [http://dx.doi.org/10.1016/j.ijforecast.2006.03.005]

[28] E.S. Gardner Jr, "Exponential smoothing: The state of the art", *J. Forecast.,* vol. 4, no. 1, pp. 1-28, 1985.

[http://dx.doi.org/10.1002/for.3980040103]

[29] M. Kumar Yadav, R. Kumar, R. Kumar Ratnesh, J. Singh, R. Chandra, A. Kumar, V. Vishnoi, G. Singh, and A. Kumar Singh, "Revolutionizing Technology with Spintronics: Devices and Their Transformative Applications", *Mater. Sci. Eng. B,* vol. 303, p. 117293, 2024. [http://dx.doi.org/10.1016/j.mseb.2024.117293]

[30] R. Kumar, R.K. Ratnesh, J. Singh, A. Kumar, and R. Chandra, "IoT-Driven Experimental Framework for Advancing Electrical Impedance Tomography", *ECS J. Solid State Sci. Technol.,* vol. 13, no. 2, p. 027002, 2024. [http://dx.doi.org/10.1149/2162-8777/ad2331]

[31] M. Alzyoud, R. Alazaidah, M. Aljaidi, G. Samara, M.H. Qasem, M. Khalid, and N. Al-Shanableh, "Diagnosing diabetes mellitus using machine learning techniques", *International Journal of Data and Network Science,* vol. 8, no. 1, pp. 179-188, 2024. [http://dx.doi.org/10.5267/j.ijdns.2023.10.006]

[32] O. Adeniyi, A.S. Sadiq, P. Pillai, M. Aljaidi, and O. Kaiwartya, "Securing Mobile Edge Computing Using Hybrid Deep Learning Method", *Computers,* vol. 13, no. 1, p. 25, 2024. [http://dx.doi.org/10.3390/computers13010025]

[33] Al Daoud EAl Daoud LAsassfeh MAl-Shaikh AAl-Sherideh AAfaneh S (2024), Enhancing Cybersecurity with Transformers: Preventing Phishing Emails and Social Media Scams2024 IEEE Conference on Dependable and Secure Computing (DSC) (31-36), Online publication date: 6-No--2024. [http://dx.doi.org/10.1109/DSC63325.2024.00017]

CHAPTER 8

Health Chain: Unlocking the Potential of Blockchain in Healthcare

Spinder Kaur[1,*] and **Ravina Gill**[1]

[1] *Department of Computer Science and Engineering, Chandigarh University, Punjab, India*

Abstract: The systematic application of blockchain in the healthcare environment will likely change the conduct, retrieval, and dissemination of health records. Blockchain technology has attracted global attention as an innovation with the potential to disrupt several industries. In this research, the application of blockchain in healthcare is explored, and 'Health Chain' is positioned as an invention that addresses the challenges of existing healthcare systems. Health Chain leverage the Internet by incorporating innovative technology like blockchain for the secure storage of EHRs, management of patients' consent, enhancement of marketing channels for drug distribution, use of IoT devices for tracking, and improvement of ease of use and efficiency. This study investigates the prospects of deploying this technology, highlighting the development of a recommended framework and directions for further research. It aims to assess the potential of utilizing blockchain in healthcare settings for more efficient information management, integration of multiple functions, and provision of services to patients in a more effective manner through the proposed framework.

Keywords: Blockchain, Data management, Decentralized networks, Electronic health records, Healthcare, Interoperability, Patientc, Pharmaceutical supply chains, The internet of things.

INTRODUCTION

An Overview of Blockchain Technology

Unlike other technologies, blockchain provides a new approach to handling transactions and records. Essentially, it is a cryptographic security system based on a chain of blocks containing chronologically ordered signed documents. These blocks are arranged in a way that prevents unauthorized alteration. The concept

* **Corresponding author Spinder Kaur:** Department of Computer Science and Engineering, Chandigarh University, Punjab, India; E-mail: spinder.e2662@cumail.in

Ramesh Kumar, Anupma Gupta, Manish Singla, Arjun Puri, Ashish Kumar Singh & Vipan Kumar (Eds.)

was first proposed by Satoshi Nakamoto in 2008, leading to the creation of Bitcoin. Blockchain technology quickly began finding applications in other industries. It has certain distinctive properties, often referred to as the "blockchain," such as decentralization, security, and others, which make it possible to reduce reliance on "middlemen" like banks, lawyers, or other intermediaries in certain types of transactions, as shown in Fig. (**1**).

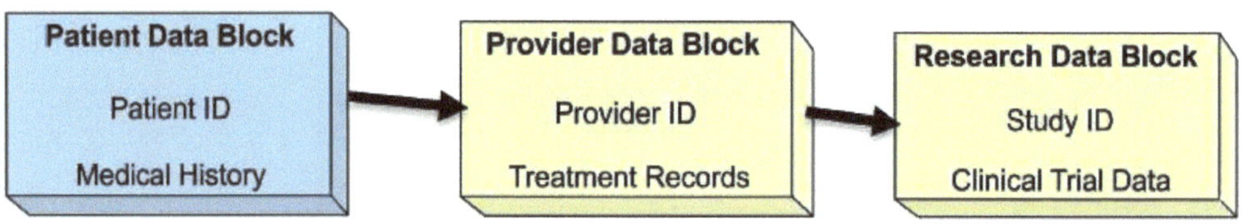

Fig. (1). Interoperability chain in healthcare data management.

The advent of blockchain technology can be described in terms of several evolutionary stages, which significantly translate into the upper layer of use for digital currencies, smart contracts, and various industries like healthcare, governance, and science, complemented by Artificial Intelligence as it develops. Operating in a distributed system, blockchain employs consensus techniques for transaction verification and cryptographic linking to secure the blockchain. This design architecture has several strengths, which enhance the security and transparency of transactions and improve the speed of processing and recording transactions. In the future, there is a high chance that blockchain will introduce radical changes in current transaction practices and data management. This technology continues to advance and shows promise of transforming many areas of activity by creating a new, effective, and safe way of communicating and exchanging information and conducting transactions.

Current Healthcare Challenges

Within the healthcare sector, a number of difficulties have arisen regarding data management and the application of data among health organizations in order to improve the quality of services rendered to patients and reduce the costs of providing those services. These difficulties include technical challenges, organizational issues, and ethical concerns. In some healthcare systems, there is no standardization of data formats, coding systems, or terminologies, which has led to the emergence of siloed health information systems and data repositories. This fragmentation makes it difficult to share information or exchange data efficiently. Additionally, fundamental aspects for smooth data exchange, such as security, privacy, and consent, are often poorly addressed, which limits the extent to which data can be shared and used.

To be sure, resource limitations, such as funding, technical manpower, and infrastructure, tend to hinder the effective use, integration, and maintenance of optimal health information systems. In addition, health professionals within the healthcare settings in the scope have a high level of illiteracy in digital devices, consequently reducing the productive use of health information systems. In more mature healthcare systems, predictive modeling of disease processes and targeted treatment with desirable effects can be implemented using techniques such as artificial intelligence, machine learning algorithms, and others. However, implementation of the proposed models is limited due to several reasons: data quality, protection of patients' privacy when using their health data for research and other purposes, and the lack of a coherent governance scheme along with proper coordination across all stakeholders involved.

Purpose of Case Study

This case study seeks to explore how this new healthcare blockchain technology has the potential to transform the sectors that use it through the innovative solution known as the Health Chain. This is because blockchain utilizes the decentralized and highly secure nature of its environment to address profound difficulties in areas such as data fragmentation, privacy concerns, and interoperability gaps within the healthcare system. It emphasizes improvements in the treatment, availability, and transfer of EHRs that blockchain technology will bring about, along with enhanced patient consent management, pharmaceutical supply chains, and even IoT devices for continuous health monitoring. All these aspects are featured in the structure and subsequent directions for further studies highlighted in the paper, showing the promise that blockchain holds for a more efficient, reliable, and patient-centric healthcare ecosystem.

BACKGROUND AND LITERATURE REVIEW

Historical Context of Healthcare IT

Health information technology (IT) has been evolving since the 1960s, when mainframe computers were first implemented to streamline hospital administration. During the 1970s and 1980s, departmental systems began to emerge, focusing on specialized applications such as laboratory information systems and radiology information systems. Early systems were very isolated and contributed little toward interoperability. The early 1990s marked a critical transitional period, with the development of electronic health records (EHRs) and hospital information systems designed to integrate separate departmental systems.

Despite these inventions, concerns such as technological limitations, high costs, and resistance from care providers constrained the proper management of data. Interoperability standards regained importance in the final years of the century, as exemplified by HL7 and federal initiatives, such as the US HITECH Act, which spurred the adoption of EHRs. With these changes, the landscape of healthcare IT faced classic challenges such as data standardization, privacy concerns, and the need for seamless information exchange across different healthcare environments, while also showing evolution in medical practices, as shown in Table **1**.

Table 1. Potential case studies in healthcare IT evolution.

Year	Milestone
1960	Installation of mainframe computers in view of their availability for the purpose of hospital billing.
1970	Development of the first EHR prototypes.
1980	Launch of PCs in health care centers.
1991	Institute of Medicine declares universal adoption of EHRs.
1996	The Health Insurance Portability and Accountability Act was enacted.
2004	Office of the National Coordinator for Health IT.
2009	HITECH Act promotes meaningful use of EHRs.
2011	Emergence of mobile health applications.
2015	Wide-scale launch of cloud-based healthcare IT solutions.
2018	Artificial intelligence/machine learning integration in healthcare analytics.
2020	Telemedicine was expected to have a tremendous boom due to the COVID-19 pandemic.
2023	Blockchain technology will be largely implemented for the secure transfer of health data.

Literature Review

Since the invention of blockchain in 2008 as the basis of Bitcoin, this technology has garnered significant attention, especially in the healthcare industry. Many longstanding concerns in the industry seem to find solutions in this decentralized, transparent, and secure technology. Various systematic literature reviews have explored the implementation of blockchain in the healthcare domain. Hölbl *et al.* (2018) utilized bibliometric methods to describe the building blocks of blockchain and trends in healthcare research. Angraal *et al.* (2017) described different types of platforms created for the implementation of blockchain in the healthcare sector. Agbo *et al.* (2019) presented examples of adoption, challenges, and possible solutions.

In addition, recent research has introduced more applications of blockchain in the health sector:

Management of Electronic Health Records

Sudeep Tanwar *et al.* (2019) [5] suggested a modern EHR system using Hyperledger Fabric, Hyperledger Composer, and Docker Container to enhance security, interoperability, and management of health records. Zhang *et al.* (2019) [6] designed authentication mechanisms that verify users and enable the exchange of medical records on the blockchain network, with a strong focus on data security and privacy through blockchain-friendly authentication schemes.

Health Information Exchange

Recently, S. Xia *et al.* (2019) [7] proposed a novel approach for safe and efficient health care data sharing, using blockchain and cloud services to transfer data between different providers.

MedChain

B. Shen *et al.* (2019) [8] developed MedChain, a blockchain-based platform for managing healthcare information. MedChain addresses some of the significant problems in healthcare data management, including security and interoperability. Kumar *et al.* (2018) [9] presents some of the key requirements and challenges in adopting blockchain technology in healthcare, such as interoperability standards, regulatory frameworks, and security.

Kassab *et al.* (2019) [10] discuss future research directions in scalability, interoperability, and data privacy to guide researchers and practitioners on blockchain applications in healthcare. While current literature highlights the promise of blockchain, it also notes limitations and overly optimistic proposals. Meinert *et al.* (2019) [11] emphasized that "practically its implementation needs careful consideration" when recommending blockchain as a technology for managing sensitive data in healthcare, medical research, and insurance.

Blockchain in Other Industries

In general, finance and supply chain management are among the most salient uses of blockchain technology. The use of blockchain in finance [16, 18] has, in fact, altered the nature of transactions to an unprecedented level of security, transparency, and efficiency [21]. Cryptocurrencies, such as Bitcoin and Ethereum, have demonstrated how blockchain, without intermediaries, reduces costs and increases the speed of transactions. In supply chain management, blockchain technology enhances traceability and, as pointed out [27, 35], transparency, allowing stakeholders to trace the origin and journey of products with precision. These applications have been highly resourceful in providing lessons about the potential of blockchain in managing complex data securely and transparently (Fig. **2**). A deeper understanding of these industries reveals the technology's potential in ensuring data integrity, automated functioning, and progressively increasing stakeholder confidence, all of which have direct applicability in the healthcare scenario. Blockchain adoption could help overcome severe problems of data fragmentation, privacy, and interoperability, leading to improved patient care and operational efficiency in health sectors.

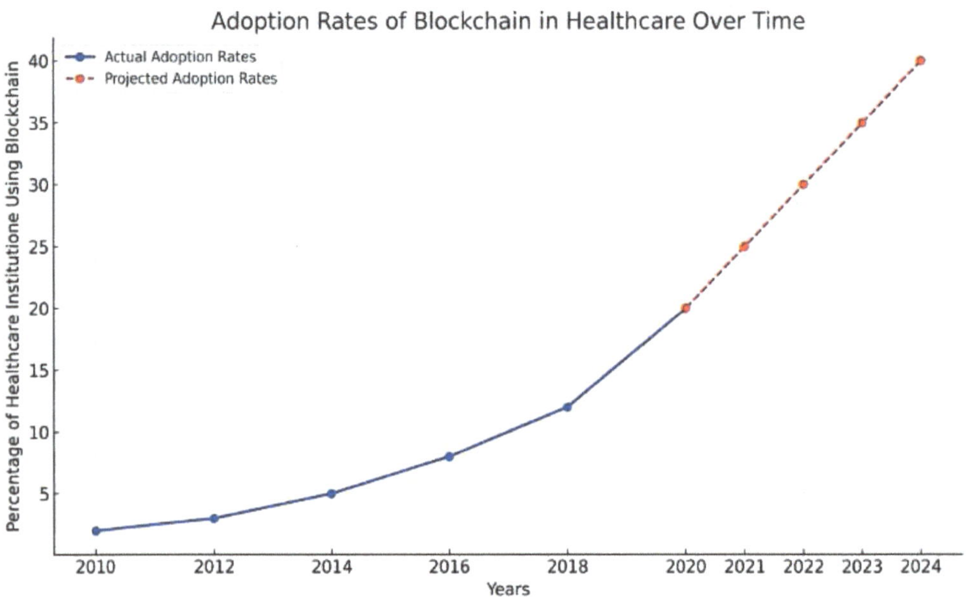

Fig. (2). Trends in blockchain adoption rates in healthcare.

HEALTH CHAIN FRAMEWORK

Technical Architecture

It is a new type of Internet trust and security because decentralization distributes power among all the nodes, eliminating a single point of failure and making it difficult to take over through attacks, as shown in Fig. (**3**) [1]. The consensus mechanism ensures agreement among participants in such networks through protocols like Bitcoin's Proof of Work (PoW) [1] or Ethereum's Proof of Stake (PoS) [2]. Other new protocols contributing to improvements include Proof of History (PoH) and Proof of Authority (PoA) for finance and supply chain management [3]. In digital identity management, blockchain enhances security and interoperability [3]. In healthcare, it extends the management of information and security through automated processes with minimal or no paperwork, advancing patient care [4]. Additionally, it can authenticate medical devices and enable real-time monitoring by integrating with IoT [5]. As healthcare moves toward digitization, smart contracts can manage clinical trials, handle patient consent, and redesign telemedicine billing, thereby improving operational efficiency and empowering patients to control their health data [6].

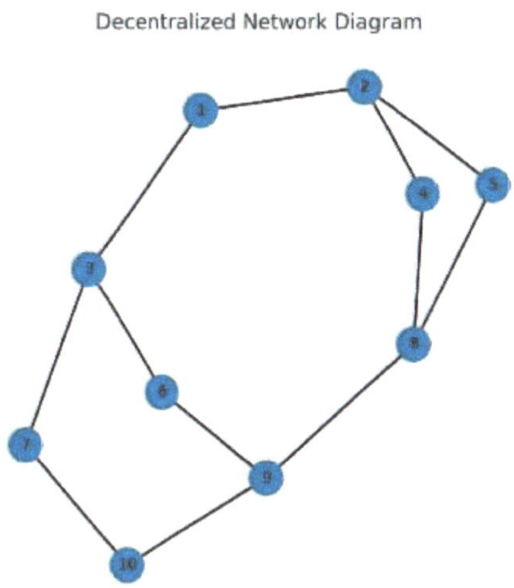

Fig. (3). Basic decentralized network structure.

Data Security and Privacy

Among the three core concepts in healthcare biotechnology, encryption, immutability, and patient consent consistently serve as the basic pillars of blockchain technology [7]. These components address major issues in data security and integrity, as well as patients' control over information in the ecosystem of digital health and care applications [8]. Integrating advanced encryption methods with the distributed architecture of blockchain enables a secure environment for sensitive medical information, preventing unauthorized access and other cyber threats [9]. One of the main benefits of the blockchain ledger is that healthcare records are permanent and cannot be altered once recorded, providing a clear record to uphold accountability and safeguard the trust of other healthcare stakeholders [10].

A particularly important advantage is the leverage this provides patients who use smart contracts introduced in third-party applications for consent management

[11]. This allows a significant shift in access control, enabling individuals to grant or withdraw permissions as ongoing research or treatment requires, in line with the principles of informed consent in modern medicine [12]. Integrating these elements through blockchain technology not only improves health data security and compliance but also strengthens the patient's role, ensuring they are active participants rather than passive custodians of their healthcare information.

Healthcare regulation compliance, for example, HIPAA or GDPR (as shown in Table **2**), is becoming increasingly important in healthcare management, as it represents more than a legal requirement [14]. It reflects a commitment to patient confidentiality, safety, and the responsible provision of health services, especially in the context of the sector's increasing digitization [15]. The focus on the privacy of health data is strictly observed in HIPAA, while GDPR, on the other hand, has no jurisdictional limitation; any organization that interacts with the information of EU citizens is bound by it [16]. Compliance can be particularly challenging for healthcare providers, especially if the organization operates offshore, as it must satisfy the requirements of both regulations [17]. A reliable implementation process requires significant commitment, including robust data protection measures, comprehensive employee education, regular evaluations and audits, and the integration of a privacy-by-design approach in all healthcare information technologies [18]. As the healthcare sector continues to embed new technologies such as telemedicine, AI diagnosis, and connected medical devices, compliance remains a persistent challenge [19]. Healthcare organizations must not only be proactive in meeting existing legislative requirements but also anticipate future demands, creating conditions for a stable organizational climate of data protection that supports continuous improvement of analytical biomarkers and patient privacy environments [20]. Such intervention measures go beyond mere compliance, potentially reducing litigation risks and enhancing overall trust in healthcare services.

Table 2. Key compliance requirements comparison: HIPAA *vs.* GDPR.

Requirement	HIPAA	GDPR
Scope	Applies to healthcare providers, health plans, and healthcare clearinghouses in the US.	This requirement applies to any organization that processes personal data of EU citizens.
Data Protection	Protects individually identifiable health information (PHI).	Protects personal data and privacy.
Consent	Requires consent for use/disclosure of PHI, with some exceptions.	Explicit consent for data processing is mandatory, though certain exceptions apply.

(Table 2) cont.....

Requirement	HIPAA	GDPR
Data Subject Rights	Limited rights (*e.g.*, access to PHI, request for amendment).	Extensive rights (*e.g.*, access, rectification, erasure, data portability).
Data Minimization	Not explicitly required, but implied through privacy and security rules.	Explicitly required: only collects data necessary for specified purpose.
Penalties	Civil and criminal penalties, up to $1.5 million per violation.	Administrative fines up to €20 million or 4% of global turnover, whichever is higher.
DataTransfer Restrictions	No specific restrictions, but must ensure PHI is protected.	Restricts data transfer outside the EU unless adequate protection is ensured.
Accountability	Requires documentation and adherence to privacy and security policies.	Requires accountability measures such as data protection officers (DPOs) and impact assessments.

Interoperability and Integration

Interoperability and integration have become critical components in the progression of blockchain technology. As the Web3 ecosystem seeks to support and work with multiple chains and layer solutions, there is an increased demand for interoperability, where different networks can communicate [23]. This interconnectedness extends beyond asset movement, building a global digital marketplace where data, smart contracts, and dApps can work and communicate across different blockchains [24]. The challenge remains in creating efficient solutions that guarantee the security and trust of individual chains while enabling effective interaction between them [25]. New technologies, such as cross-chain messaging protocols and advanced bridge architectures, are facilitating coordination among blockchains [26]. This leapfrogging technology can unlock vast potential and broaden opportunities for decentralized finance, gaming, and identity, including increased user adoption by accelerating the development of decentralized networks that were previously limited by stEXS blockchains [27]. Overall, real-time self-custodied wallets and no-code interactions are becoming increasingly feasible.

CASE STUDIES AND USE CASES

Patient-Centric Data Ownership

A new concept in managing health information is known as patient-centered data ownership. This allows individuals to own their health data [1]. This represents a 180-degree shift from the historical arrangement, where healthcare providers or organizations played a central role in maintaining the security of medical information [2]. In this model, patients can manage their health information, generate and share derived data with their healthcare providers, and become active participants in their healthcare [3]. However, this practice raises multiple questions, including how secure that information will be against unauthorized or illegal access, the ability to share and continually update such information, and how to meet both individual and public needs [4]. Moreover, as more healthcare systems become patient-oriented and data-driven, patient-centric data ownership is a key issue around which organizational, legal, and ethical frameworks must be structured to ensure equitable distribution of health information while maintaining confidentiality [5].

Clinical Trial Management

Modern Clinical Trials Management involves an intricate network of stakeholders, processes, and technologies in advancing medical knowledge responsibly and safely while ensuring patient safety and data integrity [6]. It is a discipline that must balance the worlds of science, regulatory authority, and operational feasibility [7]. As clinical research evolves, strategies that trial managers are increasingly embracing include distributed models of clinical trials, AI in patient recruitment and monitoring, and blockchain technology for enhanced data security and transparency [8]. These approaches improve operational flow in clinical trials and address challenges such as patient dropout, protocol compliance, and difficulties in distributing results [9]. However, they also introduce new ethical considerations and call for adaptive regulatory policy measures [10]. Effective Clinical Trials Management now takes a holistic approach, integrating advanced tools with robust governance structures to ensure the scientific integrity and ethical conduct of trials, while also making them accessible to patients worldwide [11]. This shift in management practices is therefore imperative for accelerating the discovery of therapies that save human lives, while maintaining the highest integrity of research and protection of participants [12].

Supply Chain Transparency

Supply chain transparency is essential in the pharmaceutical and medical device sectors to protect patients, avoid regulatory issues, and maintain operational efficiency [13]. This approach involves tracing a product throughout its supply chain journey, from raw material procurement to delivery to the patient. Advanced technologies supporting this process include blockchain, IoT, and AI-powered analytics [14].

Transparency requirements, especially amid global health crises and the circulation of counterfeit products, must be addressed to ensure trust and integrity in healthcare [15]. Strong traceability systems allow stakeholders to track the flow of goods, respond to risks, authenticate products, and monitor actual stock levels [16]. These systems also expedite recalls, enhance regulatory compliance, and generate data crucial for continuous improvement [17]. With complex global supply networks and increasing scrutiny, organizations that prioritize transparency are best positioned to address these challenges, drive innovation, and deliver safer, more reliable healthcare products to patients [18].

Telemedicine and Remote Monitoring

In this regard, the delivery mode of healthcare has transformed into telemedicine and remote monitoring through the extension of care beyond limited acute clinical environments [19]. It largely depends on the intensive use of advanced telecommunication and digital health technologies that enable real-time communication between service providers and patients [20]. However, these systems are not immune to issues regarding data security, technological obstacles, and clinical workflow adaptations [22]. Telemedicine and remote monitoring have emerged as promising solutions for healthcare systems that, in resource-constrained settings, face rising demand, improved accessibility, and the need to optimize patient outcomes and resources [23]. These technologies should be integrated in a fully complementary way, as healthcare professionals, technologists, and policymakers must work together to maximize benefits while addressing ethical concerns and human values involved in patient care [24]. The most significant technological advancements come from artificial intelligence and machine learning, which enable predictive analytics and more proactive healthcare management [25].

BENEFITS AND CHALLENGES

Transformational Benefits

The execution of clinical transformation initiatives in healthcare settings is beneficial in many dimensions, including but not limited to operational efficiency [1]. Transformational improvement in healthcare delivery can be realized by healthcare providers through adopting value metrics, streamlining processes around patients, and implementing appropriate management structures and attitudes [2].

Sophisticated approaches like these not only enable the adoption and implementation of robust management strategies to improve financial results significantly through cost savings, but also begin to change the culture of the organization [3]. They contribute to operational effectiveness, enable better clinical decisions, and encourage best practices that enhance the quality of care patients receive [4]. In addition, they help healthcare providers optimize processes, reduce fatigue, and increase job satisfaction and retention [5].

Such improvements allow hospitals to provide more services, increase stakeholders' satisfaction, and develop robust systems that facilitate adaptation to changes characterizing today's healthcare environment, as shown in Fig. (**4**) [6].

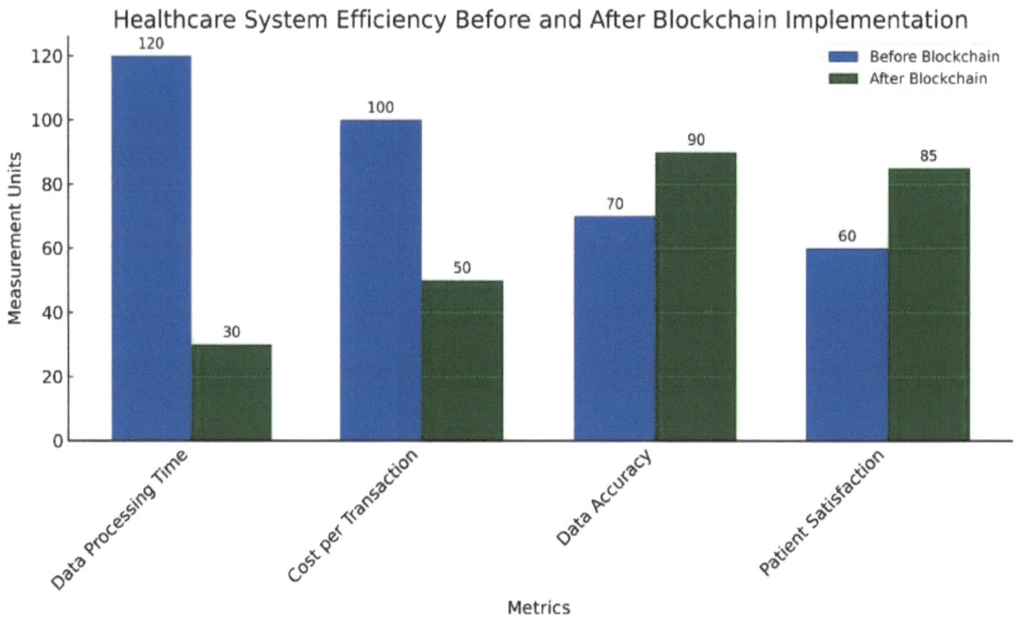

Fig. (4). Comparison chart for pre- and post-implementation metrics to efficiency in healthcare systems using blockchain.

Potential Challenges

Potential challenges include the need to address a set of complex issues that digital health innovation and transformative healthcare strategies call for [7]. For example, although the promise of these has always been exciting in attempts to improve patient outcomes as well as optimize operational efficiencies, their widespread adoption faces numerous obstacles [8].

Technical challenges include interoperability and the need for strong data security [9]. Other sources of uncertainty arise from regulatory complexities, particularly within developing regions with their transitional governance frameworks. Ethical concerns regarding what would protect patient confidentiality in this new digital age of communication and what would constitute legitimate informed consent are also relatively high [11]. Financial constraints in searching for sustainable funding models remain a persistent challenge, especially in resource-constrained environments [12]. It also involves overcoming resistance from health workers to new changes, which requires the implementation of comprehensive training programs to ensure that the use of such technologies is maximized. Implementation research needs to balance scientific rigor with pragmatic decision-making processes, which adds another layer of complexity [14].

It is only through joint, multidisciplinary efforts that stakeholders at all levels—from the policymaker at the very top down to the frontline healthcare provider—can successfully navigate the complex digital health terrain of implementation and unlock its potential.

Addressing Challenges

In the complex world of stakeholder management and risk mitigation, organizations face a myriad of challenges that require inventiveness [16]. The changed perception of risk, from being a purely negative concept to one that encompasses both opportunities and threats, constitutes a more sensitive challenge in resolving these issues [17]. It is, therefore, indispensable that organizations improve their approaches to dealing with these hindrances by utilizing the latest technological advancements, including AI-based risk assessment tools, which should be robust enough to enhance engagement with stakeholders. This calls for a change at all levels in the organization's culture, behaviors, and mindsets. Creating an atmosphere that fosters open communication, continuous learning, and adaptability will support the steps necessary to transform installation into the true implementation of change initiatives. This integrated approach enables organizations to face the increasingly shifting dynamics of the global business environment with greater resilience and agility, transforming challenges into opportunities for positive transformation and sustainable success.

STRATEGIC RECOMMENDATIONS

Roadmap for Implementation

Innovation in this application of blockchain technology should be provided to the organization in a strategic, multi-phased manner based on what the organization needs and wants. Often this process starts with a feasibility study and development of a business case, conceptualizing the blockchain network, and scoping the project. Later, at project maturity, a proof of concept can be extremely useful for pilot evaluation of ideas before committing to full-scale design and development. Designing constitutes feature road-mapping, architectural planning, and UX considerations. This occurs before entering the highly intensive development phase, which might be as short as a few months but may take up to 1 year, depending on the complexity of the solution. Deployment and system integration, followed by continuous support, require flexibility at every step of this journey so that there can be a culture of continuous learning and improvement to facilitate the smooth sailing of this shifting landscape of blockchain. The strategic recommendations are as follows:

1. Initial Assessment and Planning:
   i. Conduct a feasibility study
   ii. Identify potential use cases
   iii. Develop a business case

2. Conceptual Design
   i. Select consensus mechanism
   ii. Decide on the implementation approach

3. Proof of Concept
   i. Test core functionalities
   ii. Gather stakeholder feedback

4. Solution Design
   i. Create a detailed feature roadmap
   ii. Design system architecture
   iii. Develop UX/UI designs

5. Technology Selection
   i. Choose a blockchain platform

6. Development
   i. Set up development environment
   ii. Develop smart contracts
   iii. Build user applications
   iv. Create APIs and back-end services
   v. Conduct continuous testing

7. Preparation for Deployment
   i. Set up production infrastructure
   ii. Configure security measures

8. Pilot Deployment
   i. Deploy to a limited user group
   ii. Monitor performance
   iii. Gather and incorporate feedback

9. System Integration
   i. Ensure data consistency
   ii. Test end-to-end workflows

10. Full-Scale Deployment
    i. Roll out the solution enterprise-wide
    ii. Establish support channels

11. Post-Launch Activities
    i. Provide ongoing support
    ii. Gather and analyze usage data

12. Continuous Improvement
    i. Explore new blockchain use cases

Stakeholder Roles and Responsibilities

In the complex healthcare research and delivery context, stakeholders play multiple roles that span beyond traditional boundaries. Patients and families, who were once passive recipients, now act in both the design and execution phases of research, providing input that helps shape study protocols and increases enrollment rates. Healthcare providers work in dual roles, as caregivers and research partners, bridging clinical practice and scientific pursuit. Government agencies and policymakers operate between regulation and innovation, while pharmaceutical companies and research institutions collaborate to advance the field of medicine. Insurers and payers influence healthcare economically: their choices determine who has access to healthcare and how accessible it should be. Organizations representing communities and advocacy groups for less-heard voices ensure diversity of input for future agendas. There is a dynamic web of engagement in which relationships range from consultative to shared leadership, interweaving the roles of stakeholders. Under this intricate web of responsibilities and relationships, stakeholder management requires a nuanced approach that appreciates what each party brings to the table within a collaborative framework, developing environments responsive to the dynamic landscape of patient-centered outcomes research, as shown in Table **3**.

Table 3. Roles and responsibilities of stakeholders.

Stakeholder	Roles and Responsibilities
Patients	Providing consent, accessing personal health records
Healthcare Providers	Recording and accessing patient data, ensuring data integrity
Insurers	Accessing records for claims processing, maintaining data security
Regulators	Ensuring compliance with laws and setting standards.

Policy and Regulatory Framework

This framework of policy and regulation of health and social care services forms an intricate, complex system for achieving quality, safety, and accountability. It includes many elements, from legal mandates and trade standards to a variety of self-regulatory mechanisms that help define the compliance terrain. It is essentially a dynamic interface between the evidence base and the real world, changing constantly in response to new challenges and technological improvements. It operates at many levels, from international guidelines down to local regulations, making governance multi-layered. It establishes standards, mechanisms for monitoring, enforcement, and continuous improvement.

These include risk management protocols, reporting requirements, and audit processes aimed at developing a culture of compliance and quality improvement. This framework is not static; it continually evolves by assimilating new research findings, societal expectations, and implementation experiences. Its effectiveness depends on active engagement from all sides—from policymakers and regulators to service providers and end-users—in an ongoing cycle of evaluation, feedback, and refinement. Through this cooperative endeavor, the regulatory framework continues to be forward-looking and well-calibrated with the overarching objective of enhancing health outcomes and welfare.

FUTURE DIRECTIONS

Improving Interoperability and Integration of Data

Blockchain technology can be advanced toward full interoperability for usage within healthcare systems. This is a significant area for standard-setting and protocols that may one day be followed by healthcare providers, systems, and devices in freely and seamlessly transferring data. This would make the entire healthcare ecosystem cohesive and integrated so that all patient data is accessible, accurate, and up to date at any point across their care.

Smart Patient Consent and Data Privacy

Blockchain would enable patients to have better control over their health data, making it easier for them to be informed about the handling of their personal data. Future work may involve designing smart contracts with dynamic consent, allowing patients to give, update, or withdraw access in real time.

Blockchain-Enabled Telemedicine and Remote Monitoring

Blockchain can enhance telemedicine and remote monitoring due to the security and integrity of transmitted data. Future advances may focus on integrating blockchain with IoT devices and telehealth platforms for real-time data exchange and monitoring.

Future Research Directions

- Investigating the use of blockchain for real-time patient monitoring and alerts.
- Enhancing the integration of telemedicine platforms with blockchain for secure consultations.

Decentralized Clinical Trials

Clinical trials can benefit from blockchain technology because decentralized trials ensure data integrity and transparency for patients. Future studies will focus on developing a blockchain-based clinical trial platform that covers all stages, from volunteer recruitment to data collection and analysis, as illustrated in Fig. (**5**).

Pharmaceutical Supply Chain Management

Blockchain can lead to a significant reduction in the origin of counterfeit drugs and improve drug safety by enhancing the transparency and traceability of pharmaceutical supply chains. Its future evolution may include combining blockchain with AI and machine learning to better predict and manage risk along the supply chain.

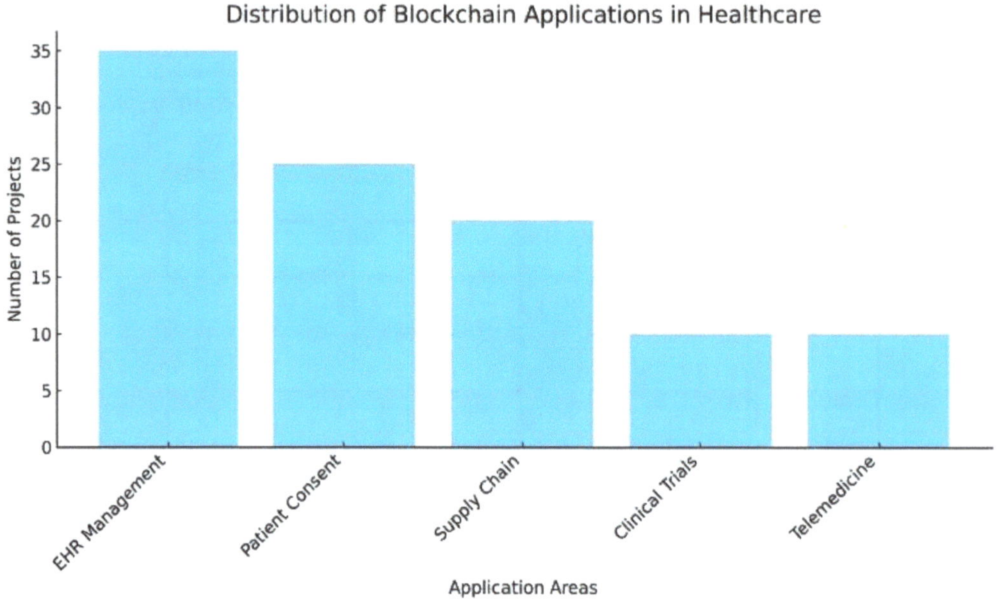

Fig. (5). Distribution of blockchain applications in healthcare.

Future Research Directions:

- Full end-to-end traceability of pharmaceutical supply chains must be developed using blockchain-based solutions.
- Exploring the integration of AI and blockchain for predictive management.
- Examine blockchain-based technologies that validate the authenticity of drugs.

Blockchain data management and security applications may be among the most important for genomic data in personalized medicine. In future work, blockchain solutions may assist in the efficient management and securing of sensitive genetic information, ensuring this information is used to provide personalized treatment plans and biomedical applications in imaging and other medical areas [36 - 39]. Blockchain and artificial intelligence might open new dimensions in healthcare. Some possible fields for future research include blockchain-AI integration for improved health data security, diagnosis, and treatment planning.

Future Research Directions:

- The integration of blockchain with AI to ensure safe and error-free diagnostics.
- Blockchain-based platforms for health analytics based on AI.
- Development of blockchain frameworks to enhance security and reliability in healthcare AI models.

CONCLUSION

Blockchain technology is a transformative opportunity in the health sector, as it delivers effective solutions to complex, long-standing problems. Interoperability and integrated data across different platforms contribute to a cohesive health system where patient information is always available and accurate. Advanced consent mechanisms, along with robust privacy protocols, empower individuals, instilling clarity and trust in the management of their health data. Blockchain-based telemedicine and remote monitoring ensure the secure flow of real-time data and improve patient care.

Decentralized clinical trials hosted on blockchain platforms are expected to revolutionize research integrity and patient engagement. In pharmaceutical supply chains, blockchain applications increase transparency, thereby reducing the risk of counterfeit products and ensuring the safety of drugs. In personalized medicine, blockchain allows for secure management of sensitive genomic information while supporting the adoption of tailored treatments.

The potential for blockchain in the healthcare sector remains vast and promising. Ongoing research into interoperability, patient consent, telemedicine, clinical trials, supply chain management, personalized medicine, AI integration, and global health data collaboration is essential to unlock the full potential of blockchain. In conclusion, blockchain is set to transform healthcare: it will make processes more efficient, secure, and effective worldwide.

REFERENCES

[1] S. Nakamoto, "Bitcoin: A peer-to-peer electronic cash system", Bitcoin.org, 2008.

[2] M. Hölbl, M. Kompara, A. Kamišalić, and L. Nemec Zlatolas, "A Systematic Review of the Use of Blockchain in Healthcare", *Symmetry (Basel),* vol. 10, no. 10, p. 470, 2018. [http://dx.doi.org/10.3390/sym10100470]

[3] S. Angraal, H.M. Krumholz, and W.L. Schulz, "Blockchain Technology", *Circ. Cardiovasc. Qual. Outcomes,* vol. 10, no. 9, p. e003800, 2017. [http://dx.doi.org/10.1161/CIRCOUTCOMES.117.003800] [PMID: 28912202]

[4] C.C. Agbo, Q.H. Mahmoud, and J.M. Eklund, "Blockchain Technology in Healthcare: A Systematic Review", *Healthcare (Basel),* vol. 7, no. 2, p. 56, 2019. [http://dx.doi.org/10.3390/healthcare7020056] [PMID: 30987333]

[5] S. Tanwar, K. Parekh, and R. Evans, "Blockchain-based electronic healthcare record system for healthcare 4.0 applications", *Journal of Information Security and Applications,* vol. 50, p. 102407, 2020. [http://dx.doi.org/10.1016/j.jisa.2019.102407]

[6] P. Zhang, J. White, D.C. Schmidt, G. Lenz, and S.T. Rosenbloom, "FHIRChain: Applying Blockchain to Securely and Scalably Share Clinical Data", *Comput. Struct. Biotechnol. J.,* vol. 16, pp. 267-278, 2018. [http://dx.doi.org/10.1016/j.csbj.2018.07.004] [PMID: 30108685]

[7] Q. Xia, E. Sifah, A. Smahi, S. Amofa, and X. Zhang, "BBDS: Blockchain-Based Data Sharing for Electronic Medical Records in Cloud Environments", *Information (Basel),* vol. 8, no. 2, p. 44, 2017. [http://dx.doi.org/10.3390/info8020044]

[8] B. Shen, J. Guo, and Y. Yang, "MedChain: Efficient Healthcare Data Sharing via Blockchain", *Appl. Sci. (Basel),* vol. 9, no. 6, p. 1207, 2019. [http://dx.doi.org/10.3390/app9061207]

[9] V. Ramani, I. Ahmad, A. Braeken, E. Harjula, and M. Ylianttila, Kumar, "Blockchain utilization in healthcare: key requirements and challenges", in 2018 IEEE 20th International Conference on e-Health Networking, Applications and Services (Healthcom), 2018, pp. 1-7.

[10] M. Kassab, J. DeFranco, T. Malas, G. Destefanis, and V.V.G. Neto, "Exploring Research Directions in Blockchain for Healthcare: A Systematic Review", *IEEE/ACM 1st International Workshop on Software Engineering for Healthcare (SEH),* 2019pp. 1-8

[11] E. Meinert, A. Alturkistani, K.A. Foley, T. Osama, J. Car, A. Majeed, M. Van Velthoven, G. Wells, and D. Brindley, "Blockchain Implementation in Health Care: Protocol for a Systematic Review", *JMIR Res. Protoc.,* vol. 8, no. 2, p. e10994, 2019. [http://dx.doi.org/10.2196/10994] [PMID: 30735146]

[12] A. Azaria, A. Ekblaw, T. Vieira, and A. Lippman, "MedRec: Using Blockchain for Medical Data Access and Permission Management", in 2016 2nd International Conference on Open and Big Data (OBD), 2016, pp. 25-30. [http://dx.doi.org/10.1109/OBD.2016.11]

[13] T.T. Kuo, H.E. Kim, and L. Ohno-Machado, "Blockchain distributed ledger technologies for biomedical and health care applications", *J. Am. Med. Inform. Assoc.,* vol. 24, no. 6, pp. 1211-1220, 2017. [http://dx.doi.org/10.1093/jamia/ocx068] [PMID: 29016974]

[14] L.A. Linn, and M.B. Koo, "Blockchain for health data and its potential use in health IT and health care related research", *ONC/NIST Use of Blockchain for Healthcare and Research Workshop,* 2016pp. 1-10

[15] X. Yue, H. Wang, D. Jin, M. Li, and W. Jiang, "Healthcare data gateways: found healthcare intelligence on blockchain with novel privacy risk control", *J. Med. Syst.,* vol. 40, no. 10, p. 218, 2016. [http://dx.doi.org/10.1007/s10916-016-0574-6] [PMID: 27565509]

[16] C. Esposito, A. De Santis, G. Tortora, H. Chang, and K.K.R. Choo, "Blockchain: A Panacea for Healthcare Cloud-Based Data Security and Privacy?", *IEEE Cloud Computing,* vol. 5, no. 1, pp. 31-37, 2018. [http://dx.doi.org/10.1109/MCC.2018.011791712]

[17] P. Mamoshina, L. Ojomoko, Y. Yanovich, A. Ostrovski, A. Botezatu, P. Prikhodko, E. Izumchenko, A. Aliper, K. Romantsov, A. Zhebrak, I.O. Ogu, and A. Zhavoronkov, "Converging blockchain and next-generation artificial intelligence technologies to decentralize and accelerate biomedical research and healthcare", *Oncotarget,* vol. 9, no. 5, pp. 5665-5690, 2018. [http://dx.doi.org/10.18632/oncotarget.22345] [PMID: 29464026]

[18] W.J. Gordon, and C. Catalini, "Blockchain Technology for Healthcare: Facilitating the Transition to Patient-Driven Interoperability", *Comput. Struct. Biotechnol. J.,* vol. 16, pp. 224-230, 2018. [http://dx.doi.org/10.1016/j.csbj.2018.06.003] [PMID: 30069284]

[19] I. Radanović, and R. Likić, "Opportunities for Use of Blockchain Technology in Medicine", *Appl. Health Econ. Health Policy,* vol. 16, no. 5, pp. 583-590, 2018. [http://dx.doi.org/10.1007/s40258-018-0412-8] [PMID: 30022440]

[20] P. Zhang, D.C. Schmidt, J. White, and G. Lenz, "Blockchain Technology Use Cases in Healthcare", *Adv. Comput.,* vol. 111, pp. 1-41, 2018. [http://dx.doi.org/10.1016/bs.adcom.2018.03.006]

[21] T.K. Mackey, T.T. Kuo, B. Gummadi, K.A. Clauson, G. Church, D. Grishin, K. Obbad, R. Barkovich,

and M. Palombini, "'Fit-for-purpose?' – challenges and opportunities for applications of blockchain technology in the future of healthcare", *BMC Med.,* vol. 17, no. 1, p. 68, 2019. [http://dx.doi.org/10.1186/s12916-019-1296-7] [PMID: 30914045]

[22] T. McGhin, K.K.R. Choo, C.Z. Liu, and D. He, "Blockchain in healthcare applications: Research challenges and opportunities", *J. Netw. Comput. Appl.,* vol. 135, pp. 62-75, 2019. [http://dx.doi.org/10.1016/j.jnca.2019.02.027]

[23] A.A. Vazirani, O. O'Donoghue, D. Brindley, and E. Meinert, "Implementing Blockchains for Efficient Health Care: Systematic Review", *J. Med. Internet Res.,* vol. 21, no. 2, p. e12439, 2019. [http://dx.doi.org/10.2196/12439] [PMID: 30747714]

[24] M. Mettler, "Blockchain technology in healthcare: The revolution starts here", in 2016 IEEE 18th International Conference on e-Health Networking, Applications and Services (Healthcom), 2016, pp. 1-3. [http://dx.doi.org/10.1109/HealthCom.2016.7749510]

[25] A. Dubovitskaya, Z. Xu, S. Ryu, M. Schumacher, and F. Wang, "Secure and Trustable Electronic Medical Records Sharing using Blockchain", *AMIA Annu. Symp. Proc.,* vol. 2017, pp. 650-659, 2018. [PMID: 29854130]

[26] K.N. Griggs, O. Ossipova, C.P. Kohlios, A.N. Baccarini, E.A. Howson, and T. Hayajneh, "Healthcare Blockchain System Using Smart Contracts for Secure Automated Remote Patient Monitoring", *J. Med. Syst.,* vol. 42, no. 7, p. 130, 2018. [http://dx.doi.org/10.1007/s10916-018-0982-x] [PMID: 29876661]

[27] C. Brodersen, B. Kalis, C. Leong, E. Mitchell, E. Pupo, and A. Truscott, *Blockchain: Securing a New Health Interoperability Experience.* Accenture LLP, 2016.

[28] R. Kumar, R.K. Ratnesh, J. Singh, A. Kumar, and R. Chandra, "IoT-Driven Experimental Framework for Advancing Electrical Impedance Tomography", *ECS J. Solid State Sci. Technol.,* vol. 13, no. 2, p. 027002, 2024. [http://dx.doi.org/10.1149/2162-8777/ad2331]

[29] M.A. Engelhardt, "Hitching Healthcare to the Chain: An Introduction to Blockchain Technology in the Healthcare Sector", *Technol. Innov. Manag. Rev.,* vol. 7, no. 10, pp. 22-34, 2017. [http://dx.doi.org/10.22215/timreview/1111]

[30] J.H. Tseng, Y.C. Liao, B. Chong, and S. Liao, "Governance on the Drug Supply Chain via Gcoin Blockchain", *Int. J. Environ. Res. Public Health,* vol. 15, no. 6, p. 1055, 2018. [http://dx.doi.org/10.3390/ijerph15061055] [PMID: 29882861]

[31] M. Benchoufi, and P. Ravaud, "Blockchain technology for improving clinical research quality", *Trials,* vol. 18, no. 1, p. 335, 2017. [http://dx.doi.org/10.1186/s13063-017-2035-z] [PMID: 28724395]

[32] A. Roehrs, C.A. da Costa, and R. da Rosa Righi, "OmniPHR: A distributed architecture model to integrate personal health records", *J. Biomed. Inform.,* vol. 71, pp. 70-81, 2017. [http://dx.doi.org/10.1016/j.jbi.2017.05.012] [PMID: 28545835]

[33] D.V. Dimitrov, "Blockchain Applications for Healthcare Data Management", *Healthc. Inform. Res.,* vol. 25, no. 1, pp. 51-56, 2019. [http://dx.doi.org/10.4258/hir.2019.25.1.51] [PMID: 30788182]

[34] A. Ekblaw, A. Azaria, J.D. Halamka, and A. Lippman, "A Case Study for Blockchain in Healthcare: "MedRec" prototype for electronic health records and medical research data", in Proceedings of IEEE Open & Big Data Conference, 2016, p. 13.

[35] D. Ichikawa, M. Kashiyama, and T. Ueno, "Tamper-Resistant Mobile Health Using Blockchain Technology", *JMIR Mhealth Uhealth,* vol. 5, no. 7, p. e111, 2017. [http://dx.doi.org/10.2196/mhealth.7938] [PMID: 28747296]

[36] M. Verma, P. Dutta, A. Basak and Ramesh. Kumar, "A Comparative Study of Material Impact on

Tunnel Field-Effect Transistor (TFET) Performance," 2024 International Conference on Emerging Trends in Networks and Computer Communications (ETNCC), Windhoek, Namibia, 2024, pp. 472-477.
[http://dx.doi.org/10.1109/ETNCC63262.2024.10767569]

[37] R. Kumar, R.K. Ratnesh, J. Singh, R. Chandra, G. Singh, and V. Vishnoi, "Recent Prospects of Medical Imaging and Sensing Technologies Based on Electrical Impedance Data Acquisition System", *J. Electrochem. Soc.,* vol. 170, no. 11, p. 117507, 2023.
[http://dx.doi.org/10.1149/1945-7111/ad050f]

[38] M. Kumar Yadav, R. Kumar, R. Kumar Ratnesh, J. Singh, R. Chandra, A. Kumar, V. Vishnoi, G. Singh, and A. Kumar Singh, "Revolutionizing Technology with Spintronics: Devices and Their Transformative Applications", *Mater. Sci. Eng. B,* vol. 303, p. 117293, 2024.
[http://dx.doi.org/10.1016/j.mseb.2024.117293]

[39] R.K. Ratnesh, M.K. Singh, V. Kumar, S. Singh, R. Chandra, M. Singh, and J. Singh, "Mango Leaves (Mangifera indica)-Derived Highly Florescent Green Graphene Quantum Dot Nanoprobes for Enhanced On–Off Dual Detection of Cholesterol and Fe 2+ Ions Based on Molecular Logic Operation", *ACS Appl. Bio Mater.,* vol. 7, no. 7, pp. 4417-4426, 2024.
[http://dx.doi.org/10.1021/acsabm.4c00292] [PMID: 38875229]

CHAPTER 9

Physico-Chemical and Spectroscopic Analysis of Municipal Solid Waste Compost in Pathankot, Punjab

Anchal Sharma[1,*], **Arjan Singh**[2] and **Vipan Kumar**[3,4]

[1] *Department of Civil Engineering, SSGI Pathankot, Punjab, India*

[2] *Department of Mechanical Engineering, CTIEMT, Punjab, India*

[3] *Department of Electronics & Communication Engineering, Sri Sai College of Engineering & Technology, Punjab, India*

[4] *Applied Science Research Center, Applied Science Private University, Amman-11937, Jordan*

Abstract: Indian cities nowadays have very poor waste management. Rapid industrialization has caused the intensification of Municipal Solid Waste (MSW) at a very progressive rate. Alternative solutions are required because disposing of waste material in open landfills and using only incineration techniques increases environmental damage. The present study conducted the physico-chemical and spectroscopic characterization of municipal solid waste compost in Pathankot city of the Punjab region to analyze its appropriateness for various purposes. The compost was prepared collectively from the college hostel mess waste and from the dry leaves available in the proximity of the college campus. The compost pit was constructed with dimensions of 4*4 feet in the college campus area. The study revealed the examination of nutrient concentration and structural behaviour during the entire process of waste degradation. The study concluded that the region's waste is rich in organic materials and, hence, composting is an appropriate option for processing and minimizing waste.

Keywords: Compost pit, Compost, Pathankot, Physico-chemical, Punjab, Solid waste, Waste characterization.

INTRODUCTION

The current scenario of solid waste management in India is highly inadequate [1, 2]. According to reports, 1,35,198 tons of municipal solid waste are produced daily (TPD) in India [3]. However, only around 70% of this waste is collected,

* **Corresponding author Anchal Sharma:** Department of Civil Engineering, SSGI Pathankot, Punjab, India; E-mail: anchalsharma881@gmail.com

Ramesh Kumar, Anupma Gupta, Manish Singla, Arjun Puri, Ashish Kumar Singh & Vipan Kumar (Eds.)

and the proportion of treated waste is reported to be about 12%. The main factor behind the lower collection efficiency is the inadequate availability of transportation vehicles [4, 5]. The economic conditions of cities and towns in any state significantly affect waste generation [6, 7]. Currently, Indian cities produce between 0.4 and 0.7 kilograms of garbage per person per day [8], making it extremely difficult to manage solid waste generation in most cities [9]. Poor municipal solid waste management, particularly open dumping, remains a major challenge in developing nations like India [10 - 12].

In terms of waste minimization, composting is one of the easiest and most cost-effective ways to reduce the burden of waste. Aerobic composting using the windrow method is documented as a cost-effective technique that can be employed as a soil amendment. This method is relatively simple, and its by-product significantly enhances soil fertility. Additionally, composting can complement incineration and recycling efforts, thereby reducing the overall amount of waste in urban areas [13, 14]. Furthermore, it is less expensive than other waste management options and represents an environmentally friendly alternative [15].

Zero Waste refers to designing and managing products systematically to avoid and eliminate the volume and toxicity of waste and materials [16]. It is a goal that is ethical, economical, efficient, and visionary. To achieve this, the main focus is to provide the college campus with a sufficient number of color-coded garbage bins. The process begins at the source, with mini-dustbins placed in offices and classrooms according to different types of waste. The aim is proper source segregation of municipal solid waste. Providing dustbins across the campus and establishing a composting site will help train students in waste segregation and composting practices.

The present study presents a case study of the waste management scenario at a college campus located in Pathankot city, Punjab. Punjab generates 4,376 tonnes of waste per day, with almost half treated and the remainder dumped illegally in open landfills. A similar problem of open dumping, rather than segregated disposal, was identified within the college campus. Additionally, burning of dry leaves was practiced, which can directly harm the environment and contribute significantly to air pollution. The open dumping of waste on campus is shown in Fig. (**1**).

Fig. (1). Open dumping of mixed waste.

Due to inadequate dumping of waste, groundwater, air, and land may become contaminated. To prevent this worsening situation, it is very important to ensure proper segregation of dry and wet waste and to avoid illegal burning of waste, which can release harmful gases. In this regard, the study comprised the physical characterization of solid waste generated in the area. Additionally, to assess the nutrient content and structural changes during the composting process, the study also included the physico-chemical and heavy metal characterization of compost made from garden and college mess waste in the Pathankot, Punjab, campus.

STUDY LOCATION

Pathankot is located at a latitude of 32.2733° North and a longitude of 75.6522° East, with a population of 148,937. The study location comprised the SSGI College campus area. The compost was prepared collectively from the college hostel mess waste and dry leaves from gardens within the campus.

The compost pit was constructed with dimensions of 4*4 feet. Students and coordinators were trained to process green waste into fertilizer, which can later be used in the gardens. Recyclable waste was delivered to local Kabadi Walas, while non-recyclable waste was sent to the nearest legal dumping ground in the near future.

Environmentally friendly approaches to organic waste management and recycling are urgently needed worldwide [17]. Millions of tons of organic waste are generated every day in a variety of forms, including yard waste and tree trimmings, organic residual sludge from sewage treatment plants, agricultural waste, food processing plant waste, restaurant food waste, farm animal waste, and more [18].

Alternative solutions are required since disposing of garbage in landfills, the ocean, or through incineration pollutes the environment. Cities have a unique opportunity to improve their overall waste collection services by implementing waste diversion through composting initiatives. Additionally, compared to other options such as waste-to-energy programs, composting offers a cost-effective approach to environmentally sustainable waste management. However, composting is often overlooked in integrated municipal solid waste management plans, and unfortunately, several initiatives worldwide have already failed.

The compost pit prepared on the college campus is displayed in Fig. (**2**)

Fig. (2). Compost pit prepared in a college campus.

Through microbial activity, composting is a biological decomposition process that transforms organic matter into stable, nutrient-rich compounds that can be utilized to improve and sustain soil fertility. The compost feed is shown in Fig. (**3**).

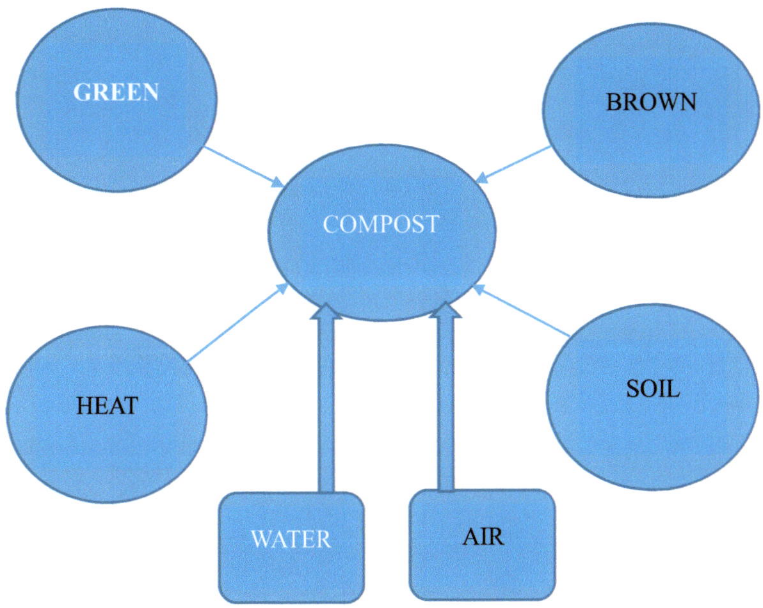

Fig. (3). Compost Feed.

The benefits of compost include providing nutrients to crops as fertilizer and acting as a soil conditioner. It also reduces expenses on commercial chemical fertilizers for recreational gardeners and commercial farmers alike. Compost can also be used for land and stream reclamation, wetland construction, and landfill cover.

MATERIAL AND METHODOLOGY

Sampling Procedure and Analysis

The physical characterization of solid waste was carried out on-site using a weighing process. The compost was prepared by collecting wet waste from the hostel mess, while dry leaves from the college campus were added as feed to the compost pit. Compost samples were collected periodically from the pit on days 20, 40, and 60 of the composting process, in accordance with USEPA Part 503 Rule requirements (USEPA 1995). During the collection of garbage from the compost pit, the temperature (°C) of the compost samples was also measured. The gravimetric method was used to determine the moisture content of the compost samples. Electrical conductivity and pH were measured using IS codes (IS 2720, Part 26, 1986) and the FAI technique (FAI 2007). The contents of organic carbon, nitrogen, and phosphorus were determined following the procedure outlined in IS-10158, 1982. Calcium and magnesium were estimated using the titration method according to the specified standards. Heavy metal concentrations were assessed using the atomic absorption spectroscopy (AAS) method as per the relevant standards.

RESULTS AND DISCUSSIONS

The physical characterization of solid waste is the primary and most important aspect of waste processing for understanding its behaviour and nature. Waste characterization helps adopt various processing methods for solid waste. In this regards, the physical characterization of waste in the study region, Pathankot, Punjab, is shown in Table **1**.

Table 1. Physical Characterization of Municipal Solid Waste in the study region.

Sr. No.	Parameters	Study Region (Pathankot)
1.	Organic waste	50.20
2.	Paper	16.35
3.	plastic	20.00
4.	Glass	2.42
5.	Metal	1.80
6.	Inert	6.00
7.	Rubber	1.23
8.	Textile	2.00

The results indicated that the largest fraction of total waste was organic waste. Plastic waste constituted the second-highest fraction, while paper waste formed the third-highest fraction of municipal solid waste in the study region of Pathankot, Punjab. The high proportion of plastic waste is attributed to the fact that plastic is not banned in Punjab, unlike in Himachal Pradesh. Paper waste forms the third-highest fraction, followed by inert waste, due to the nature of the waste sampled from the institution. The higher amount of organic matter, along with sufficient moisture content present in the solid waste [19], makes it suitable for composting, as per the NEERI report. Composting is a biological decomposition process that converts organic matter into stabilized, nutrient-rich products through microbial activity, which can be used to enrich soil fertility [20, 21]. Manufacturing compost from solid waste is a low-cost alternative to other decomposition processes such as incineration and landfilling. Compost is an excellent fertilizer because it contains essential plant nutrients such as potassium, phosphorus, and nitrogen, which improve soil quality by enhancing soil aeration [22].

The physico-chemical characterization of compost is presented in Table **2**. The compost temperature on the 60th day of the decomposition process was recorded using a thermometer and observed to be 54°C for the study region; however, it was noted as 59°C on day 20 and 55°C on day 40. The decrease in temperature over time is attributed to the exothermic reactions during the decomposition of organic matter, indicating the progressive degradation of the organic matter present in the municipal solid waste.

Table 2. Physico-chemical characterization of compost.

Sr. No.	Parameters	20th day	40th day	60th day
1.	Temperature (^{0}c)	59	55	54
2.	pH	8.02	7.63	7.17
3.	Electrical conductivity (ms/cm)	5.8	5.2	5.0
4.	Moisture content (%)	42.00	37.04	31.87
5.	Calcium (mg/l)	11.23	14.76	17.54
6.	Magnesium (mg/l)	7.92	5.04	3.89
7.	Organic carbon (%)	16.28	14.42	11.22
8.	Available nitrogen (mg/kg)	0.59	0.68	0.78
9.	Phosphorus (%)	2.11	1.34	0.81
10.	Potassium (%)	12.12	8.45	6.80
11.	C/N ratio	26.81	25.73	25.02

The pH of the compost samples from the study regions was noted as 7.17 on day 60 of the degradation process. The pattern showed alkaline to almost neutral values, revealing the decomposition of organic matter and indicating the maturity of the compost. Compost's alkaline pH can cause the formation of ammonium gas in the atmosphere, which tends to increase the number of dangerous pathogenic microorganisms.

The major plant nutrients, nitrogen, phosphorus, and potassium (NPK), are important parameters for evaluating compost quality as a fertilizer. The concentrations of nitrogen, phosphorus, and potassium were found to be 0.78 mg/Kg, 0.81 mg/Kg, and 6.80 mg/Kg, respectively, for the study region, which is considered appropriate according to the range of FCO standards stipulated by the

Ministry of Agriculture, Government of India. Additionally, parameters such as magnesium and calcium increased progressively during the degradation process, which is beneficial for compost production. Organic carbon, on the other hand, decreased with the degradation process, attributed to the release of carbon dioxide and methane. Furthermore, the compost's C/N ratio was 25.02, which is higher than the minimum requirements specified by the FCO (Fertility Control Order).

The concentration of heavy metals is an important parameter for assessing the pollution potential of compost material before its application to food crops. The heavy metal analysis of the compost material from the study region is presented in Table **3**.

Table 3. Heavy metal analysis of compost in Pathankot (Punjab).

Sr. No.	Parameter (mg/Kg)	20th day	40th day	60th day	FCO Standards
1.	Lead	15.20	10.03	9.01	100
2.	Zinc	6.94	4.08	3.74	1000
3.	Chromium	2.65	2.12	1.07	50
4.	Cadmium	1.78	1.56	1.32	5
5.	Copper	8.54	6.98	5.12	500
6.	Nickel	26.56	24.12	22.51	50

The heavy metal analysis indicated that the compost collected from the study region, Pathankot, was rich in nickel and lead compared to other metals, due to the high proportion of batteries and cells found in the dumpsite as mixed waste. However, various past studies reported a low cadmium content in compost samples, *i.e.*, 0.01 mg/L in most cities [22]. Apart from this, the concentrations of all metals were within the permissible limits of the Fertility Control Order (FCO).

CONCLUSION

The physical characterization of waste in the campus area revealed a higher proportion of organic waste, followed by plastic and paper waste. This indicates a positive potential for the composting process, provided there is adequate segregation of solid waste generated in the area. The heavy metal concentrations in the study region were found to be slightly higher than the acceptable ranges as per the Fertility Control Order guidelines set by the Indian Government's Ministry of Agriculture. In this context, the compost from the study region can be utilized as a soil conditioner in gardens and other areas, but its application in organic farming is limited. If this compost is to be used for agricultural purposes, municipal solid waste compost must be further treated to improve its quality.

Most of the physico-chemical parameters were observed within the permissible limits of the FCO. Heavy metals were higher in the initial stage of organic matter decomposition but showed a decreasing trend as the compost approached the final stage of decomposition. Aerobic composting not only reduces the weight of solid waste in dumpsites but also provides an effective method for decomposing wet waste, which can then be followed by recycling processes.

In summary, adequate solid waste segregation and compost treatment are necessary before applying compost to land for high-value food crops to enhance its suitability and value for farming. Larger color-coded bins, green for organic waste and blue for dry paper or plastic waste, are one suggested method to achieve this. This system makes it easy to identify and process waste for various applications. Materials from the blue bins are segregated and sent to the Waste Management Plant (Pathankot) for recycling, while materials from the green bins are fed directly into the composting pits.

A campaign has been launched on campus to strongly discourage the use of plastic bags and to promote the adoption of personalized cloth and paper bags provided to the Institute community for shopping, as part of the ongoing effort to become a no-plastic campus. Plastic bottles have drastically decreased as a result of the campus's “Green Bottle Campaign,” which encourages the use of reusable bottles for water and other beverages. Implementing zero-waste management will minimize discharges to land, water, and air, thereby reducing environmental threats. In a nutshell, the SSGI college campus is on track to become the first “zero-waste” campus in Pathankot city, provided proper waste segregation is followed, and adequate provisions are made to manage solid and liquid waste, ensuring a cleaner environment.

REFERENCES

[1] S.O. Adefemi, and E.E. Awokunmi, "The impact of municipal solid waste disposal in Ado Ekiti metropolis, Ekiti State, Nigeria", *J. Environ. Sci. Technol.*, vol. 3, pp. 186-189, 2009.

[2] S. Bruun, T.L. Hansen, T.H. Christensen, J. Magid, and L.S. Jensen, "Application of processed organic municipal solid waste on agricultural land – a scenario analysis", *Environ. Model. Assess.*, vol. 11, no. 3, pp. 251-265, 2006. [http://dx.doi.org/10.1007/s10666-005-9028-0]

[3] P.S. Bundela, S.P. Gautam, A.K. Pandey, M.K. Awasthi, and S. Sarsaiya, "Municipal solid waste management in Indian cities-A review", *Int. J. Environ. Sci.*, vol. 1, p. 4, 2010.

[4] *Urban Waste Profile. Parivesh Bhavan, Delhi. Central Pollution Control Board, Ministry of Environment and Forests, Government of India, New Delhi,* 2010.http://cpcb.nic

[5] G. Gomez, M. Meneses, L. Ballinas, and F. Castells, "Characterization of urban solid waste in Chihuahua, Mexico", *Waste Manag.*, vol. 28, no. 12, pp. 2465-2471, 2008.

[http://dx.doi.org/10.1016/j.wasman.2007.10.023] [PMID: 18248808]

[6] B. Gupta, and S.K. Arora, "Municipal solid waste management in Delhi – the Capital of India", *Int. J. Innov. Res. Sci. Eng. Technol.,* vol. 5, no. 4, pp. 5130-5138, 2016.

[7] A. Hanc, P. Novak, M. Dvorak, J. Habart, and P. Svehla, "Composition and parameters of household bio-waste in four seasons", *Waste Manag.,* vol. 31, no. 7, pp. 1450-1460, 2011. [http://dx.doi.org/10.1016/j.wasman.2011.02.016] [PMID: 21419614]

[8] F. Ingelmo, M.J. Molina, M.D. Soriano, A. Gallardo, and L. Lapeña, "Influence of organic matter transformations on the bioavailability of heavy metals in a sludge based compost", *J. Environ. Manage.,* vol. 95, suppl. Suppl., pp. S104-S109, 2012. [http://dx.doi.org/10.1016/j.jenvman.2011.04.015] [PMID: 21570172]

[9] S. Kumar, J.K. Bhattacharyya, A.N. Vaidya, T. Chakrabarti, S. Devotta, and A.B. Akolkar, "Assessment of the status of municipal solid waste management in metro cities, state capitals, class I cities, and class II towns in India: An insight", *Waste Manag.,* vol. 29, no. 2, pp. 883-895, 2009. [http://dx.doi.org/10.1016/j.wasman.2008.04.011] [PMID: 18595684]

[10] B.M. Kiran, and S. Srikantaswamy, "Heavy Metals Assessment in Municipal Solid Waste Dumpiste, Mysore, Karnataka, India", *International Journal of Advanced Research in Engineering and Technology,* vol. 5, no. 4, pp. 88-93, 2014.

[11] S.E. Mbuligwe, G.R. Kassenga, M.E. Kaseva, and E.J. Chaggu, "Potential and constraints of composting domestic solid waste in developing countries: findings from a pilot study in Dar es Salaam, Tanzania", *Resour. Conserv. Recycling,* vol. 36, no. 1, pp. 45-59, 2002. [http://dx.doi.org/10.1016/S0921-3449(02)00009-5]

[12] B. Manohara, "Study of Decomposition Pattern During Aerobic Composting of Municipal Solid Waste by Physico-chemical and Spectroscopic Method", *Int. J. Chemtech Res.,* vol. 10, no. 10, pp. 27-34, 2017.

[13] P. Mandal, M.K. Chaturvedi, J.K. Bassin, A.N. Vaidya, and R.K. Gupta, "Qualitative assessment of municipal solid waste compost by indexing method", *Int. J. Recycl. Org. Waste Agric.,* vol. 3, no. 4, pp. 133-139, 2014. [http://dx.doi.org/10.1007/s40093-014-0075-x]

[14] *Assessment of Quality of Compost from Waste Processing Plants of Calicut Delhi and Jalandhar.* Environment Engg Res Inst: Nagpur, 2009.

[15] A. Puri, M. Kumar, and E. Johal, "Solid-waste management in Jalandhar city and its impact on community health", *Indian J. Occup. Environ. Med.,* vol. 12, no. 2, pp. 76-81, 2008. [http://dx.doi.org/10.4103/0019-5278.43265] [PMID: 20040983]

[16] R. Rana, R. Ganguly, and A.K. Gupta, "Physico-chemical characterization of solid waste management from tricity region of northern India", *J. Mater. Cycles Waste Manag.,* vol. 20, no. 1, pp. 678-689, 2018. [http://dx.doi.org/10.1007/s10163-017-0615-3]

[17] M. Sharholy, K. Ahmad, G. Mahmood, and R.C. Trivedi, "Municipal solid waste management in Indian cities – A review", *Waste Manag.,* vol. 28, no. 2, pp. 459-467, 2008. [http://dx.doi.org/10.1016/j.wasman.2007.02.008] [PMID: 17433664]

[18] A.V. Shekdar, "Sustainable solid waste management: An integrated approach for Asian countries", *Waste Manag.,* vol. 29, no. 4, pp. 1438-1448, 2009. [http://dx.doi.org/10.1016/j.wasman.2008.08.025] [PMID: 19081236]

[19] J.K. Saha, N. Panwar, and M.V. Singh, "An assessment of municipal solid waste compost quality produced in different cities of India in the perspective of developing quality control indices", *Waste Manag.,* vol. 30, no. 2, pp. 192-201, 2010. [http://dx.doi.org/10.1016/j.wasman.2009.09.041] [PMID: 19857948]

[20] S. Sethi, N.C. Kothiyal, A.K. Nema, and M.K. Kaushik, "Characterization of Municipal Solid Waste

in Jalandhar City, Punjab, India", *J. Hazard. Toxic Radioact. Waste,* vol. 17, no. 2, pp. 97-106, 2013. [http://dx.doi.org/10.1061/(ASCE)HZ.2153-5515.0000156]

[21] S. Srivastva, V. Krishna, and I. Sonkar, "Characterisation and management of municipal solid waste: a case study of Varanasi city, India", *Int. J. Curr. Res. Acad. Rev.,* vol. 2, no. 8, pp. 10-16, 2014.

[22] A. Sharma, R. Ganguly, and A.K. Gupta, *Matrix method for evaluation of existing solid waste management system in Himachal Pradesh, India,* 2018. [http://dx.doi.org/10.1007/s10163-018-0703-z]

CHAPTER 10

Assessment of Groundwater Indices and Physico-Chemical Characterization of Leachate in Pathankot, Punjab

Anchal Sharma[1,*], **Arjan Singh**[2] and **Vipan Kumar**[3,4]

[1] *Department of Civil Engineering, SSGI Pathankot, Punjab, India*

[2] *Department of Mechanical Engineering, CTIEMT, Punjab, India*

[3] *Department of Electronics & Communication Engineering, Sri Sai College of Engineering & Technology, Punjab, India*

[4] *Applied Science Research Center, Applied Science Private University, Amman-11937, Jordan*

Abstract: Open dumpsites are one of the major causes of environmental degradation, and their impact on groundwater largely depends on leachate concentration. The environment can be severely polluted by leachate percolation if it is not properly managed. The present study examines the effect of leachate percolating into aquifers and soil in the vicinity of dumping sites in the study region of Pathankot, Punjab. The study primarily focuses on the evaluation of groundwater characterization, leachate characterization, and heavy metal analysis for quality assessment. In this context, the leachate pollution index is used to categorize the toxic potential of leachate so that immediate remedial actions can be implemented at dumping locations. Higher concentrations of leachate pollutants increase the likelihood of toxicity and, in turn, the pollution of groundwater and soil. From this perspective, groundwater pollution is an impending environmental issue that requires urgent consideration. In a nutshell, the current study endorses the construction of engineered landfills and effective leachate management to ensure the sustainability and conservation of groundwater aquifers and soil quality.

Keywords: Groundwater, Leachate, Pathankot, Physico-chemical characterization, Pollution index, Punjab, Water quality.

INTRODUCTION

Open dumping practices in Indian cities are recognized as one of the major threats and critical issues to groundwater reserves. This problem is prominent not only in

* **Corresponding author Anchal Sharma:** Department of Civil Engineering, SSGI Pathankot, Punjab, India; E-mail: anchalsharma881@gmail.com

Ramesh Kumar, Anupma Gupta, Manish Singla, Arjun Puri, Ashish Kumar Singh & Vipan Kumar (Eds.)

India but throughout the world (United States Environmental Protection Agency, US EPA). It has been observed from the literature that more than 80% of solid waste is directly abandoned on the ground in an illegal manner [1, 2]. During rainfall, dumped solid waste comes into contact with water, and the decomposed water in the form of leachate infiltrates into the ground and hence pollutes the groundwater [3]. This leachate percolates through soil voids and contaminates groundwater aquifers to a greater extent [4, 5]. The level of groundwater pollution due to filtration of leachate is influenced by many factors, such as rainfall, depth, and the physical and chemical composition of solid waste [6]. However, sometimes the pollutants may not exceed drinking water limits, but still there may be a significant threat to human health and the environment [7 - 9]. The leachate pollution index and water quality index have been evaluated by many researchers to categorize the toxic potential of leachate so that immediate corrective actions can be executed at the dumping location [10, 11]. Higher values of the leachate pollution index indicate a greater probability of leachate toxicity and, in turn, pollution of groundwater and soil [12 - 14].

The environment can be eminently polluted by means of leachate percolation if not handled properly [15, 16]. The present evaluation study comprised the influence of leachate percolated into the aquifers and soil near the vicinity of dumping sites in the study region of Pathankot, Punjab. The study focused on the evaluation of groundwater characterization, heavy metal analysis, and leachate characterization for the quality assessment of leachate and groundwater.

From this viewpoint, groundwater pollution is an impending environmental issue and needs to be addressed. The water quality index is a tool to evaluate the quality of water and soil, which is substantially beneficial for policymakers for the ecological and efficient supervision of groundwater reserves [17 - 21]. In a nutshell, the study recommends the construction of engineered landfills and proper management of leachate so that groundwater reserves and soil quality may be maintained and preserved.

STUDY LOCATION

Pathankot is located at 32.2733° N (Latitude) and 75.6522° E (Longitude) and has a population of 148,937. The study area includes a large dumpsite near the city. The dumpsite receives waste from Pathankot on a daily basis. The current scenario of the dumpsite near Pathankot city is shown in Fig. (**1**).

Fig. (1). Current Scenario of the dumpsite near Pathankot City (Study location).

MATERIAL AND METHODOLOGY

The preliminary visit and survey of the dumpsite were conducted on a frequent basis. The physical characterization of municipal solid waste was analyzed in the dumpsite area. The analysis and literature study revealed that the study region is rich in organic waste, followed by plastic waste, as plastic is not banned in Punjab; hence, a huge amount of plastic waste was observed in the dumpsite, which may be a source of many harmful pollutants. The physico-chemical characterization of groundwater was analyzed to determine the possible contamination levels due to the percolation of leachate. The pre-survey of the dumpsite is shown in Fig. (**2**).

Fig. (2). Preliminary visit and survey of dumpsite.

The groundwater samples were collected from submersible and hand pumps located downstream at various distances of 1, 2, 2.5, and 3 km from the dumpsites. The samples were collected periodically across different seasons. The first round of sampling was conducted during the peak summer season (April 2023), followed by sampling in the rainy and winter seasons of the year. The sample collection from the dumpsite is shown in Fig. (**3**).

Fig. (3). Sample collection of groundwater near the dumpsite area.

The groundwater samples were collected from the vicinity of the dumpsite at distances of 1 km, 2 km, 2.5 km, and 3 km. The sources of groundwater sample collection were tubewells, submersible pumps, and handpumps. Additional samples were collected farther from the dumpsite at distances of 7 km, 8 km, 9 km, and 10 km to examine the comparative analysis of pollution potential.

The physico-chemical parameters, including pH, turbidity, nitrates, BOD, COD, magnesium chloride, *etc.*, were analyzed in accordance with APHA 2012 [22 - 24]. The nitrate concentration, specifically, was measured using a UV-VIS Spectrophotometer 108 (APHA 2012).

RESULTS AND DISCUSSIONS

Groundwater Characterization

The physico-chemical characterization of the groundwater samples for three seasons at different distances was investigated for the study region, Pathankot, and is shown in Table **1**. The pH level was found to be in the neutral range and within the limits prescribed by BIS. The TDS content of groundwater in the study region was observed to be on the higher side, as per the specifications given by BIS, ranging between 490 mg/l and 521.7 mg/l. The electrical conductivity of groundwater samples varied between 427.20–453.08 μS/cm for the specified dumpsite in the study area of Pathankot city, Punjab, indicating a higher ionic concentration in the groundwater due to heavy percolation of leachate.

Table 1. Physico-chemical characterization of groundwater.

Sr. No.	Parameters	Distance (Km)	S1	S2	S3
1.	pH	1	7.38	7.45	7.68
		2	7.32	7.42	7.61
		2.5	7.26	7.37	7.59
		3	7.20	7.34	7.52
2.	Turbidity	1	7	8	8
		2	5	7	7
		2.5	4	6	6
		3	3	4	4
3.	TDS	1	490.2	498.4	521.7
		2	483.4	493.6	513.8
		2.5	481.6	487.8	510.2
		3	479.8	480.2	504.4
4.	Phosphate	1	0.043	0.058	0.063
		2	0.042	0.052	0.060
		2.5	0.037	0.045	0.056
		3	0.032	0.042	0.051
5.	BOD	1	0.37	0.39	0.42
		2	0.35	0.36	0.40
		2.5	0.30	0.32	0.35
		3	0.29	0.30	0.32

(Table 1) cont.....

Sr. No.	Parameters	Distance (Km)	S1	S2	S3
6.	COD	1	3.6	3.8	4.12
		2	3.5	3.6	4.06
		2.5	3.0	3.2	3.98
		3	2.8	2.9	3.75
7.	Electrical Conductivity	1	427.20	440.08	453.08
		2	424.48	438.48	451.28
		2.5	402.89	432.09	449.14
		3	389.48	431.42	448.34
8.	Chlorides	1	24.12	26.50	29.53
		2	23.80	24.24	27.48
		2.5	25.12	21.09	22.13
		3	24.89	20.48	21.63
9.	Nitrates	1	7.6	7.9	8.2
		2	7.2	7.7	7.8
		2.5	6.7	6.9	7.5
		3	5.8	6.2	6.9
10.	Magnesium	1	40.24	43.08	46.48
		2	38.52	42.85	43.14
		2.5	35.43	38.10	40.24
		3	34.15	37.08	38.14

Turbidity of groundwater samples was observed within the permissible range for three seasons as per the values prescribed by BIS. Other parameters, including magnesium, chlorides, phosphates, and nitrates, were observed on the higher side but still within the permissible limits of BIS standards.

The concentration of BOD in the groundwater was 0.37 mg/l, 0.39 mg/l, and 0.42 mg/l for the three seasons, respectively, and was within the permissible limit of the Indian standard. However, it showed the presence of some fractions of dissolved organics in the groundwater. Likewise, the COD values of the groundwater samples in the study region for three seasons were 3.6 mg/l, 3.8 mg/l, and 4.12 mg/l, respectively, indicating the presence of non-biodegradable fractions in the groundwater [25 - 27].

The heavy metal analysis of the study location in Pathankot, Punjab, has been summarized in Table **2**.

Table 2. Heavy Metal analysis of groundwater samples.

Sr. No.	Parameters	Pathankot (Punjab)	Permissible Limits
1.	Cadmium	2.07±0.04	5.00
2.	Chromium	43.72±2.79	50.00
3.	Copper	57.84±1.25	300.00
4.	Iron	2576.02±36.23	-
5.	Manganese	27.12±2.89	-
6.	Nickel	41.23±0.76	50.00
7.	Lead	67.45±8.45	100.00
8.	Zinc	38.09±1.95	1000.00

The heavy metal analysis of the selected dumpsite in Pathankot city revealed that parameters including cadmium, chromium, copper, nickel, manganese, lead, iron, and zinc were within the permissible limits. However, some parameters, such as nickel and lead, were found to be on the higher side of the prescribed limits due to the extensive decomposition of batteries and plastic materials in the open dumpsite. In addition, the leachate sample was analyzed for physico-chemical characterization, and the results are summarized in Table **3**.

Table 3. Physico-chemical characterization of leachate.

Sr. No.	Parameters	Summer Season	Rainy season	Winter season
1.	pH	7.90	8.13	8.45
2.	TDS	3214	3278	3376
3.	TSS	2145	2276	2321
4.	Cl^-	675	760	823
5.	SO42-	278.5	321.7	387.5
6.	PO43-	1.24	1.37	1.53
7.	TH	645.8	720.2	739.5

The physico-chemical characterization of the leachate sample revealed that the leachate parameters exceeded the limits prescribed by Indian Standards for disposal and, when compared with previous studies [28 - 30], indicated a higher toxic level of leachate produced in the dumpsite of the study region in Pathankot city, Punjab.

CONCLUSION

The present study comprised the physico-chemical assessment and heavy metal analysis of groundwater at the selected dumpsite in Pathankot, Punjab. The study revealed that the quality of groundwater was generally in the medium range but tended toward poor quality according to Indian Standards for groundwater quality. Groundwater samples collected near the dumpsite showed higher contamination than samples collected farther away, clearly indicating that contamination occurred due to the dumping of solid waste and the percolation of leachate into groundwater reserves. With increasing distance from the dumpsite, the contamination level decreased.

Seasonal variations also affected groundwater quality, with concentrations of various parameters showing an increasing trend over time. Similarly, heavy metal analysis revealed that, although the parameters were currently within permissible limits, there is a potential for them to exceed the prescribed limits if open dumping continues at the Pathankot dumpsite. The physico-chemical concentration of leachate was found to be high and exceeded permissible standards.

In conclusion, the study indicates that open dumping of waste should be avoided, and engineered landfills with liners and leachate collection systems should be constructed to preserve groundwater aquifers and protect the environment.

CONSENT DECLARATION

Verbal and informed consent were obtained from all individuals appearing in the images included in this chapter. All participants were clearly informed about the purpose of image usage, and they voluntarily agreed to the inclusion of their photographs for academic publication.

REFERENCES

[1] P.S. Bundela, S.P. Gutam, A.K. Pandey, M.K. Awasti, and S. Sarsaiya, "Municipal solid waste management in India-A review", *Int. J. Environ. Sci.*, vol. 1, pp. 591-606, 2010.

[2] G. Bhalla, P. K. Swamee, A. Kumar, and A. Bansal, *Assessment of groundwater quality near municipal solid waste landfill by an aggregate index method,* 2010.

[3] Barjinder Bhalla, M. Saini, and M. Jha, "“Effect of age and seasonal variations on leachate characteristics ofmunicipal solid waste landfill”, IJRET", *Int. J. Res. Eng. Technol.*, vol. 2, no. 8, pp. 223-232, 2013.
[http://dx.doi.org/10.15623/ijret.2013.0208037]

[4] B. Bhalla, M.S. Saini, and M.K. Jha, *Leachate contamination potential of unlined municipal solid waste landfill sites by leachate pollution index. International Journal of Science, Environment ISSN 2278–3687 and Technology,* 2014 a.

[5] B. Bhalla, M.S. Saini, and M.K. Jha, *Assessment of municipal solid waste landfill leachate treatment*

efficiency by leachate pollution index, 2014 b.

[6] N. Dharmarathne, and J. Gunatilake, ""Leachate Characterization and Surface Groundwater Pollution at Municipal Solid Waste Landfill of Gohagoda", Sri Lanka", *International Journal of Scientific and Research Publications,* vol. 3, no. 11, 2013.

[7] I. R. Ilaboya, E.O. Oti, GO. Ekoh, L.O. Umukoro, and B.O. Enamuotor, *Assessment of water quality index of some selected boreholes around dump sites in Nigeria,* 2014.

[8] B. Jhamnani, and S.K. Singh, "Groundwater contamination due to Bhalaswa landfill site in New Delhi, India", *J. Environ. Sci. Eng.,* vol. 1, no. 3, pp. 121-125, 2009.

[9] D. Kumar, and B.J. Alappat, "Evaluating leachate contamination potential of landfill sites using leachate pollution index", *Clean Technol. Environ. Policy,* vol. 7, no. 3, pp. 190-197, 2005. [http://dx.doi.org/10.1007/s10098-004-0269-4]

[10] S. Kumar, J.K. Bhattacharyya, A.N. Vaidya, T. Chabrabarti, S. Devotta, and A.B. Akolkar, *Assessment of the status of municipal solid waste management in metro cities, state capitals, class 1 cities, and class 2 towns in India: an insigh,* 2009. [http://dx.doi.org/10.1016/j.wasman.2008.04.011]

[11] S.S. Kale, A.K. Kadam, S. Kumar, and N.J. Pawar, "Evaluating pollution potential of leachate from landfill site, from the Pune metropolitan city and its impact on shallow basaltic aquifers", *Environ. Monit. Assess.,* vol. 162, no. 1-4, pp. 327-346, 2010. [http://dx.doi.org/10.1007/s10661-009-0799-7] [PMID: 19301138]

[12] N. Kalra, R. Kumar, S.S. Yadav, and R.T. Singh, "Water quality index assessment of ground water in Koilwar block of Bhojpur (Bihar)", *J. Chem. Pharm. Res.,* vol. 4, no. 3, pp. 1782-1786, 2012.

[13] S. Kanmani, and R. Gandhimathi, "Assessment of heavy metal contamination in soil due to leachate migration from an open dumping site", *Appl. Water Sci.,* vol. 3, no. 1, pp. 193-205, 2013. [http://dx.doi.org/10.1007/s13201-012-0072-z]

[14] S. Mor, K. Ravindra, R.P. Dahiya, and A. Chandra, "Leachate characterization and assessment of groundwater pollution near municipal solid waste landfill site", *Environ. Monit. Assess.,* vol. 118, no. 1-3, pp. 435-456, 2006. [http://dx.doi.org/10.1007/s10661-006-1505-7] [PMID: 16897556]

[15] A. Puri, M. Kumar, and E. Johal, "Solid-waste management in Jalandhar city and its impact on community health", *Indian J. Occup. Environ. Med.,* vol. 12, no. 2, pp. 76-81, 2008. [http://dx.doi.org/10.4103/0019-5278.43265] [PMID: 20040983]

[16] K. Parameswari, B.V. Mudgal, and P. Nelliyat, "Evaluation of groundwater contamination and its impact: an interdisciplinary approach", *Environ. Dev. Sustain.,* vol. 14, no. 5, pp. 725-744, 2012. [http://dx.doi.org/10.1007/s10668-012-9349-5]

[17] N. Raman, and D.S. Narayanan, "Impact of solid waste effect on ground water and soil quality nearer to pallavaram solid waste landfill site in Chennai", *Rasayan J. Chem.,* vol. 1, no. 4, pp. 828-836, 2008.

[18] S. Renou, J.G. Givaudan, S. Poulain, F. Dirassouyan, and P. Moulin, "Landfill leachate treatment: Review and opportunity", *J. Hazard. Mater.,* vol. 150, no. 3, pp. 468-493, 2008. [http://dx.doi.org/10.1016/j.jhazmat.2007.09.077] [PMID: 17997033]

[19] I.M. Rafizul, M. Alamgir, and M.M. Islam, "Evaluation of contamination potential of sanitary landfill Lysimeter using leachate pollution index", *Thirteenth International Waste Management and Landfill Symposium,* 2011 Sardinia

[20] R.K. Ranjan, A. Ramanathan, P. Parthasarathy, and A. Kumar, "Hydrochemical characteristics of groundwater in the plains of Phalgu River in Gaya, Bihar, India", *Arab. J. Geosci.,* vol. 6, no. 9, pp. 3257-3267, 2013. [http://dx.doi.org/10.1007/s12517-012-0599-1]

[21] P.K. Swamee, and A. Tyagi, "Describing water quality with aggregate index", *J. Environ. Eng.,* vol.

126, no. 5, pp. 451-455, 2000. [http://dx.doi.org/10.1061/(ASCE)0733-9372(2000)126:5(451)]

[22] P.K. Swamee, and A. Tyagi, *Improved method for aggregation of water quality subindices,* 2007. [http://dx.doi.org/10.1061/(ASCE)0733-9372(2007)133:2(220)]

[23] R. Srivastava, V. Krishna, and I. Sonkar, "Characterization and management of municipal solid waste: a case study of Varanasi city, India", *Int. J. Curr. Res. Acad. Rev.,* vol. 2, no. 8, pp. 10-16, 2014.

[24] A. Sharma, M.K. Kaushik, A. Kumar, and A. Bansal, "Assessment of ground water quality and feasibility of tire derived aggregates for use as a leachate drainage material", *Electron. J. Geotech. Eng.,* vol. 21, pp. 3113-3136, 2016.

[25] A. Sharma, R. Ganguly, and A.K. Gupta, *Characterization of municipal solid waste in sundernagar, Himachal Pradesh, India,* 2017.

[26] A. Sharma, R. Ganguly, and A.K. Gupta, *Matrix method for evaluation of existing solid waste management system in Himachal Pradesh, India,* 2018. [http://dx.doi.org/10.1007/s10163-018-0703-z]

[27] T. Swamee, S. Sharma, P. Singh, and R. Dobhal, "Water quality assessment in terms of water quality index", *American Journal of Water Resources,* vol. 1, no. 3, pp. 34-38, 2013.

[28] M. Umar, H.A. Aziz, and M.S. Yusoff, "Variability of parameters involved in leachate pollution index and determination of LPI from four landfills in Malaysia", *Int. J. Chem. Eng.,* vol. 2010, pp. 1-6, 2010. [http://dx.doi.org/10.1155/2010/747953]

[29] *Guidelines for Drinking-water Quality THIRD EDITION INCORPORATING THE FIRST AND SECOND ADDENDA Volume 1 Recommendations,* 2008.

[30] V.M. Wagh, D.B. Panaskar, A.M. Varade, S.V. Mukate, S.K. Gaikwad, R.S. Pawar, A.A. Muley, and M.L. Aamalawar, "Major ion chemistry and quality assessment of the groundwater resources of Nanded tehsil, a part of southeast Deccan Volcanic Province, Maharashtra, India", *Environ. Earth Sci.,* vol. 75, no. 21, p. 1418, 2016. [http://dx.doi.org/10.1007/s12665-016-6212-2]

SUBJECT INDEX

A

B

C

D

Q

R

S

T

www.ingramcontent.com/pod-product-compliance
Lightning Source LLC
LaVergne TN
LVHW070950210926
848482LV00002B/166

* 9 7 9 8 8 9 8 8 1 4 5 8 8 *